R-3661-FF/CSTP

Redesigning Teacher Education

Opening the Door for New Recruits to Science and Mathematics Teaching

Linda Darling-Hammond, Lisa Hudson,
Sheila Nataraj Kirby

March 1989

Supported by
The Ford Foundation

Center for the Study of
the Teaching Profession

JAN 1 5 1990

370
.732
0913
D221

FOREWORD

Currently, most teacher education programs—preservice preparation for teaching—are embedded in four-year undergraduate colleges. The programs are structured to allow students to satisfy the college's requirements for an undergraduate degree while meeting existing state requirements for teacher certification. The programs work reasonably well for those who decide before college or during their freshman or sophomore years that they wish to become teachers.

However, not everyone interested in teaching makes that early commitment. Some will decide as they approach the end of college or shortly thereafter. Others will decide in the middle of a career. Still others will decide after they have completed a career.

Fortunately for these potential recruits to teaching, the demand for knowledgeable teachers of science and mathematics (and certain other fields) is present and growing. The supply of traditional school of education graduates can and must be augmented to meet the demand for knowledgeable science and mathematics teachers. However, traditional teacher preparation is not structured to allow access to these latecomers.

This report examines recent programmatic responses to breaking down institutional barriers. In most instances examined, special programs were designed to prepare mathematics and science teachers and each program was targeted on a specific population such as retired persons, persons changing careers, or recent college graduates.

Many programs quickly found, however, that to survive they needed to recruit students from a variety of sources; some needed to expand beyond mathematics and science. Others, located outside universities or outside traditional university structures, had to become part of the colleges of education. In other words, they needed to become more general, more traditional, more university-like.

Creating focused, targeted programs is not enough. Potential recruits to teaching need to become aware of their existence. Marketing and advertising is important but, if programs that accommodate nontraditional students are to become a dependable source of teachers, they must be institutionalized. In some measure, new recruits will be attracted to teaching in proportion to their knowledge of the existence of means by which to become teachers. They will decide to become teachers, in part, because they know that they can become teachers, because programs are available.

The problems identified in this report underscore the importance of unbundling teacher education from the undergraduate curriculum. By unbundling teacher education, universities can create and market teacher preparation in all teaching fields to prospective teachers of all ages and career stages. Moving teacher education to the graduate-level, a move urged by many recent reports on education, would have this effect.

Enrollments in existing graduate programs in teacher education have been growing rapidly in recent years, suggesting that new recruits can be attracted to serious graduate study. Moreover, graduate programs have been able to attract a variety and quality of candidates that undergraduate programs do not. These include: a larger representation of minority groups (whose enrollment in undergraduate programs is actually dwindling); persons who bring a wide range of occupational experience; and persons who have developed an interest in teaching a subject on which they have become expert.

A major problem is, of course, cost, including opportunity cost. Programs can be tailored to meet the needs of those who are employed. Scholarships and fellowships can be provided. Industries reducing their staffs can provide financial aid as an outplacement service. Industries concerned about the quality of public education can provide financial aid as an early retirement incentive.

Redesigning teacher education by moving it to the graduate level will open the door to teaching to a wider variety of persons than does the existing undergraduate arrangement. Schools currently have difficulty finding sufficient numbers of knowledgeable science and mathematics teachers. Moreover, there are spot shortages of teachers in other fields. Structuring teacher education so that knowledgeable persons of all ages and career stages can enter the field will create additional supplies of teachers who know their subjects well and have had the opportunity to learn how to teach before they are given the responsibility to teach. As the reader will learn in this report, new recruits to teaching value most those programs that had sufficient scope, depth, and duration to prepare them for their important and challenging new responsibilities.

>Arthur E. Wise
>Center for the Study of
>the Teaching Profession

Ministry of Education, Ontario
Information Services & Resources Unit,
13th Floor, Mowat Block, Queen's Park,
Toronto   M7A 1L2

PREFACE

This report, along with two companion notes, presents the results of an eighteen-month study sponsored by The Ford Foundation; the study examined new recruits into elementary and secondary mathematics and science teaching and the programs that prepare them as teachers. These "new recruits" are not undergraduate teacher education majors; they are from "nontraditional" pools such as retirees, homemakers, career switchers, and recent college graduates with mathematics and science degrees. For the purposes of this study, the definition of "nontraditional" is extended to include teachers in other fields seeking retraining as mathematics and science teachers.

An earlier Note, *Recruiting Mathematics and Science Teachers Through Nontraditional Programs: A Survey* (N-2736-FF/CSTP), presents an overview of 64 nontraditional programs in place across the country. It emphasizes recruitment, program requirements, and conditions that seem important in ensuring the viability of the programs. The present report focuses on a small sample of these nontraditional programs and their students. It describes the structure of the programs and the motivations, perceptions, and labor market experiences of the individuals who enter and graduate from these programs.

A companion Note, *Recruiting Mathematics and Science Teachers Through Nontraditional Programs: Case Studies* (N-2768-FF/CSTP), contains more detailed case studies of the nine programs we examined in this report.

This report and its companion Notes should be of interest to audiences at both the national and state level who are interested in teacher preparation as well as issues of teacher supply and demand. They should also be of interest to those concerned with improving teacher education or implementing similar nontraditional programs.

These data should help policymakers and practitioners consider how such programs can best be designed and targeted to potential teachers to produce a greater supply of well-qualified mathematics and science teachers for our nation's schools.

SUMMARY

Over the last several years, there has been growing concern about shortages of qualified mathematics and science teachers, which would threaten the quality of education in these fields. Although teacher retirements and increased course requirements are boosting demand for mathematics and science teachers, the number of newly trained entrants graduating from undergraduate teaching programs declined by more than one-half throughout the 1970s and early 1980s, falling far short of the number of new teachers needed.

In recognition of the increased demand for teachers and the perceived inability of traditional undergraduate teacher training programs to meet this need, several policy initiatives have been implemented. One of these initiatives is the development of postgraduate-level programs to train "nontraditional" recruits for mathematics and science teaching. These recruits include recent college graduates with degrees in mathematics or science; individuals in science-related fields who are retiring or who want to make a midcareer switch to teaching; and teachers who initially prepared to teach in areas other than mathematics or science.

Nontraditional preparation programs aim to entice individuals into mathematics or science teaching who might not otherwise be willing or able to enter. They strive to reduce the various costs of entry—financial costs, transaction costs, or opportunity costs—so that individuals in other occupations or with different training will find teacher preparation feasible. This report describes a number of these special teacher preparation programs and the nontraditional recruits they prepare for elementary or secondary mathematics or science teaching. The programs are classified into three categories:

- *Nontraditional recruitment programs* are designed to provide potential teachers from nontraditional pools with the coursework and other requirements for full certification in mathematics or science. These programs do not require changes in state policies regarding teacher training or licensure.
- *Alternative certification programs* are designed to increase the potential supply of teachers by preparing them to meet revised state certification requirements for entering teaching.
- *Retraining programs* are designed to help teachers already trained in other fields to obtain endorsement or certification in mathematics or science.

What sets nontraditional teacher preparation programs apart from previous attempts to resolve teacher shortages is that they seek to find a compromise between competing demands for quality and for quantity. That is, they attempt to find, recruit, and prepare for teaching careers individuals who were not attracted to traditional undergraduate preparation programs, while maintaining the requirement that these recruits be certified to teach their subjects. Whereas most responses to teacher shortages have either relied on efforts to attract recruits to traditional teacher education programs, or skirted these programs entirely by allowing the hiring of uncertified entrants, nontraditional initiatives seek to modify the characteristics of traditional teacher education programs that may serve to limit the numbers and quality of individuals they attract.

PROGRAM CHARACTERISTICS

Although nontraditional programs all strive to reduce or overcome some of the potential barriers to entry into teaching, they do so in very different ways. Program duration, intensity, and content vary tremendously, as do financial aid availability and the prospect of quickly assumed paid employment. For example, the extent of required study varies substantially across programs, ranging from as few as nine course credits to as many as 45, conducted over as little as 16 weeks to as much as two or more years, completed before entry into teaching or after a full-time job has been assumed, managed in concert with a guided practicum or without any practicum or student teaching experience at all.

In spite of this diversity, current nontraditional teacher preparation programs typically share a number of common features as well. In accordance with their goal of providing a more feasible alternative to traditional undergraduate programs, these programs tend to target coursework more closely on recruits' certification needs and are less costly in time and money than traditional programs. Focusing on one particular type of recruit—retirees, midcareer transfers, or former teachers, for example—allows a program to create a curriculum tailored specifically to the needs and prior training of that group.

Nontraditional preparation programs also tend to exist in locations where teacher demand is high and a viable recruitment pool is easily accessible. Even given a "high supply" location, though, these programs usually must actively recruit to obtain a sufficient number of well-qualified participants. Programs targeted toward retirees and midcareer transfers have sometimes found that recruitment is made more difficult by teaching's relatively low salary level; many potential candidates lose interest once they discover how much a teacher earns. This appears to be especially problematic in areas where early retirees have a wide range of opportunities in private industry.

Total program costs can be substantial, ranging from $2,500 to $10,000 per recruit. Many of these costs are covered by third-party sponsors—states, school districts, federal government or foundation grants, or industry donations; as a result, some programs charge little or no tuition. For those that do, recruits take advantage of individual fellowships, grants, or loans, and many have paid teaching positions or internships while enrolled in their programs. The amount of out-of-pocket tuition costs the recruits pay varies widely by program type, with retraining programs being the least costly and midcareer programs being the most costly to participants. Not surprisingly, recruits rate procedures and arrangements that ease the financial and time burdens of teacher preparation as one of the most attractive features of nontraditional programs.

However, some of the same features that make programs attractive to potential recruits also make them vulnerable. Dependence on outside donors means that the programs may disappear when funds are scarce. Also, funding sometimes disappears when there is a perception that shortages have been "solved" or when other budgetary priorities take precedence for local or state funds. Similarly, the narrow targeting of recruitment pools that allows programs to tailor their approach for a specific type of recruit can backfire if the target pool shrinks or is not responsive to the incentives these programs offer. As a result of these kinds of problems, eight of the 64 programs we initially surveyed had been discontinued by the time the study was completed. Several others were unsure that they would continue operating in the following year; many others had substantially changed their focus, scope, or approach to preparation and recruitment.

Many programs have responded to these factors with changes that have helped them to maintain their enrollments and remain viable. In addition to expanding their recruitment pools, many programs have become institutionalized as part of their affiliated university's regular master's degree or postgraduate teacher certification programs. Some programs initially run by school districts (retraining or alternative certification programs) eventually become university-based as well. They are thus protected from some of the vicissitudes of local funding and labor market shifts, profiting as well from the faculty and programmatic resources of the university. The programs in turn strengthen the capacity of universities to develop and maintain postbaccalaureate teacher education programs, sometimes also encouraging innovation in "regular" program organization and curricula.

Judging from recruits' comments, these programs appear to be successful at meeting their basic goal: preparing nontraditional recruits to enter the classroom quickly. In so doing, however, the programs face the same conflicts as traditional teacher preparation programs, but here they are accentuated by the desire to prepare teachers in a shorter period of time. Programs must provide sufficient coursework in teaching methods, balance theory and practice, and pace instruction appropriately. The programs seem to be somewhat uneven in their ability to address these issues. The nontraditional programs that follow a more "traditional" preparation approach—providing substantial pedagogical coursework *before* recruits enter the classroom and providing supervision and graduated assumption of responsibility during a practicum—are more effective in the eyes of their participants and graduates. Programs that severely truncate coursework and place candidates in teaching positions without adequate preparation or supervision are less well-rated by recruits. Unfortunately, these include the alternative certification programs that, in our sample, trained a sizable number of new entrants to teaching.

Recruits' recommendations for program improvement focused on four ways of assisting recruits' adjustment to teaching: (1) making educational coursework more rigorous, more specific to subject matter pedagogical needs, and more practically informative; (2) providing longer, more varied, and more closely supervised teaching practicum experiences (including observation of other teachers); (3) providing better placement assistance for both the practicum and for those seeking teaching positions; and (4) providing greater access to mentor teachers or other assistance once in the classroom. The usefulness of involving expert, experienced classroom teachers as both course instructors and supervisors was also frequently noted.

Programs tend to be designed on the assumption that candidates come either with adequate subject matter background or, in the case of retraining programs, with adequate pedagogical background. In the cause of efficient preparation, they strive to provide additional coursework only in the areas not presumably already mastered. However, candidates in all types of programs often felt that they would have benefited from more coursework in both areas, filling in subject matter gaps or holes in pedagogical understanding—and in the area bridging subject matter to pedagogy, the acquisition of subject-specific teaching methods. It may be that preparation for the complexity of teaching cannot treat content and pedagogy as entirely separate areas of study.

The often expressed need for a longer or "better" practicum experience also reinforces the importance of integrating content and method. Among these nontraditional recruits, however, the two groups that do not have a student teaching/practicum experience (retrainees and alternative certification recruits who enter the classroom without prior practice teaching) also stressed the need for it. Those in retraining programs felt that the opportunity to observe and practice with current mathematics or science teachers would assist them in learning to apply their new knowledge. Alternative certification recruits felt even more strongly that classroom observation and practice before entering the classroom would improve their initial teaching performance and adjustment. In addition, alternative certification recruits keenly felt the need for more and better supervision on the job. Surprisingly, given the emphasis these programs place on on-the-job training *in lieu* of coursework, they in fact offered the least supervision of any program type, with many recruits receiving assistance only a few times during an entire year.

THE NEW RECRUITS

Our survey results show that retirees do not appear to constitute a large part of the nontraditional recruitment pool. Programs that initially focused on retirees have usually had to turn to other sources of recruits to maintain their enrollments. Homemakers also appear to participate in these programs in very small numbers. Recent college graduates, on the other hand, enter nontraditional teacher preparation programs in greater numbers, although relatively more of these recruits are likely to choose some other career upon graduation or after entering teaching temporarily.

Teachers from other subject areas are also a good pool from which to draw, although retraining programs have high attrition rates. Also, many retrained teachers do not appear to obtain mathematics or science teaching positions, at least within the first year or two. Finally, individuals who change occupations also appear to be a smaller, yet viable recruitment pool. However, only about one-third of these career switchers come from science-related occupations; among these, about half come from lower-paying technical, support, and service jobs, rather than from professional and managerial occupations.

In addition to bringing to teaching a wide range of backgrounds and experiences in scientific and nonscientific fields, bachelor's degree subject-area training, and a range of past instructional experiences, these recruits also bring more demographic diversity to the field. Overall, they are older than the "traditional" program recruit, more likely to be male than the overall teaching force (but more likely to be female than the mathematics and science teaching force), and more likely to be members of minority groups.

In spite of these differences, the recruits greatly resemble "traditional" teachers in their reasons for wanting to teach; they share an interest in subject matter, children, and contributing to society. Both traditional and nontraditional recruits appear to perceive teaching as an occupation that offers few financial rewards, but many personal and social rewards.

In fact, a fair number of our sample's "new recruits" to teaching were not in fact new entrants to the occupation. About half of the program participants were former teachers; many had already been teaching mathematics or science without certification. Of the retrainees, one-third were teaching mathematics or science before they entered a program to "retrain" (and certify) them for this task. Of the other recruit types, nearly 20 percent had taught previously, virtually all of them in mathematics and science. Most were becoming certified through alternative programs, presumably having taught for some period of time without certification. These recruits, then, had already chosen to teach; the program's role was more to help them satisfy minimal requirements than to attract them to the occupation.

PROGRAM GRADUATES

Graduates of these nontraditional programs do not have difficulty finding teaching jobs. In our sample, graduates of the recent B.A. programs, in particular, felt that they were very well-received by local districts. Among the recent B.A. and midcareer programs we examined, familiarity with the program and its graduates seemed to encourage the acceptance of new program graduates. On the other hand, for a few candidates, initial acceptance was difficult; this seemed to be especially true for alternative certification program graduates, who felt that many school personnel were reluctant to accept alternative certification as a legitimate or thorough means of teacher preparation.

When asked about their satisfaction with teaching and their current teaching assignment, these nontraditional program graduates voice concerns very similar to those expressed by teachers in general, and by new teachers in particular. Like other beginning teachers, they experience some degree of "reality shock" and find that student discipline and motivation are two of their most difficult problems. On the other hand, many graduates find their interaction with students and freedom within the classroom to be quite fulfilling. Although it is reassuring that many new recruits find teaching to be more rewarding than they had expected, the preponderance of unexpected problems and disappointments suggests that, as others have noted, teachers need a better "reality base" from which to begin their teaching careers, as well as more support within their first few years of teaching.

The question of whether and how long these new recruits plan to stay in teaching is an important one. After all, nontraditional teacher preparation programs are in some sense effective only if they prepare qualified teachers to enter *and* remain in the classroom. The program graduates in our sample entered and have remained in teaching at rates comparable to those for traditionally prepared teachers. Excluding those who were already teachers before program entry, 86 percent of program graduates entered teaching and about 75 percent were still teaching within an average of two years after program completion. Among our sample of participants and graduates, approximately 70 percent plan to remain in teaching for "a while" although only about half plan to make teaching a career.

On this measure, these programs appear to be about as successful as more traditional programs in recruiting teachers for the classroom. However, given the later career choice that many of these new recruits have made, and the time and energy that many have devoted to training for this new career, one might expect an even higher level of planned retention. It seems, however, that it is the difficult nature of teaching itself that is largely responsible for most new recruits' considerations of leaving teaching. Working conditions feature heavily in expressed dissatisfaction, especially schools' apparent emphasis on paperwork and nonteaching activities at the expense of teaching time. The relatively low pay teachers receive also features as a strong detractor. In one respect, though, program preparation may be a factor: Many recruits felt they could have been better prepared to handle classroom management and student discipline than they were. Though actual experience is a necessity in developing this type of skill, better preservice training could expand recruits' understanding of student behavior and their repertoire of management strategies.

It is likely that raising the entry and retention rates of these nontraditional recruits will involve reforms targeted at improving the working conditions that all new teachers—including this sample of nontraditional recruits—find unsatisfactory. These include: raising teacher salaries, providing first-year in-class support and less difficult initial teaching assignments, reducing course loads and bureaucratic interferences with teaching, and providing better preparation for dealing with classroom management and student motivation.

CONCLUSIONS

By the indicators available in this study, nontraditional programs appear to be supplying a nontrivial number of mathematics and science teachers to school systems in their areas. The 64 programs we surveyed in 1986–87 enrolled over 2,000 science and mathematics candidates among them, supplying perhaps as many as 10 percent of entering teachers in those fields that year. However, the rate at which programs are created, modified, and ended suggests that the share of teaching positions they help to fill will continue to be unpredictable in the immediate future.

Generally, programs that have survived and maintained their enrollments have done so by remaining flexible in how they seek funding, in who they recruit, and in how they package their program. Those that develop a reputation for producing well-qualified teachers and those that become institutionalized as part of university-based teacher preparation programs are most likely to attract and place a steady stream of recruits.

The attractions of graduate-level teacher training for these nontraditional recruits also suggest ways in which existing teacher preparation programs may be able to attract a wider range of applicants by modifying certain aspects of their programs: offering postgraduate training, more flexible scheduling, tailored coursework, support for training of like cohorts of students, and careful supervision of practicum experiences. The availability of financial aid also clearly affects program success in recruiting entrants.

This study also indicates that, for all their promise, nontraditional teacher preparation programs cannot fully overcome other attributes of teaching that make recruitment and retention of teachers difficult. In broad terms, there are two ways to increase the supply of entrants to an occupation: increase the benefits or attractions to the field or reduce the costs of entry. Efforts to improve teacher salaries and working conditions are intended to address the first of these by increasing the attractions of teaching. Nontraditional programs attend to the second of these strategies: They aim to lower the financial and opportunity costs of preparing to teach. Our results point to both the possibilities and limits of this approach to increasing supply.

On the one hand, lowering entry costs does engender interest on the part of many individuals from other fields. On the other hand, many potential candidates lose interest when they confront teaching conditions firsthand in a practicum or initial teaching experience. Given the substantial investments made in their education by governments, school districts, and foundations, these losses to teaching could be counted as debits on the policy evaluation ledger. Although not the fault of the programs, they indicate that the cost-reduction strategy for overcoming teacher shortages will prove effective only if the benefits of teaching are adequate to retain the new recruits whose customized preparation has been so generously subsidized.

ACKNOWLEDGMENTS

Any project like this one owes a great deal to the efforts of many people. We are indebted to our project officer, Barbara Scott Nelson, of The Ford Foundation for her support and interest. We are particularly grateful for the cooperation of the administrators of the selected teacher preparation programs, and for their willingness to provide us with names and addresses of their students, past and present. These administrators are: Henry Bindel, Jr. (George Mason University's Alternative Science Teacher Program), Mary Louise Ortenzo and Jay Shotel (George Washington University's Midcareer Math and Science Master of Arts in Education), Delia Stafford and Teddy McDavid (Houston Alternative Certification Program), Linda Calderon and Sheila Cassidy (Los Angeles County Mathematics and Science Teacher Retraining Program), William W. Smith (Lyndhurst Fellowship Program at the University of North Carolina at Chapel Hill), Dr. William Halpern and Patricia Wentz (Math-Science Teacher Education Program, University of West Florida), Patricia Graham and Steven K. Million (South Carolina Critical Needs Certification Program at Winthrop College), Rose Marie Smith (Texas Woman's University's THA-MASTER Program), Richard Clark, John Fischetti, and Klaus Schultz (MESTEP at the University of Massachusetts at Amherst), and Perry Phillips (West Virginia University's Post-B.A. Teacher Certification Program). We also appreciate the cooperation of Bonnie Troxell, coordinator of the Susquehanna University Teacher Intern Program, for her (and her interns') assistance in pretesting our survey instruments. We also thank all our survey respondents for their patience and cooperation.

We thank Jennifer A. Hawes, RAND's Senior Survey Director, and her staff for their assistance in designing and fielding the surveys of program participants and graduates, and Priscilla Schlegel for her assistance in database management. We are also grateful to Barnett Berry, Associate Director of the University of South Carolina's Education Policy Center, for his assistance in conducting interviews and administering surveys at Winthrop College. Our special appreciation goes to Luetta Pope, Nancy Rizor, and Linda Tanner for their excellent typing and their inexhaustible patience, and to Patricia Bedrosian for her careful, constructive editing.

All of these individuals contributed to the success of this study. None of them though, are responsible for any failings that remain. For those, the authors blame each other.

CONTENTS

FOREWORD iii

PREFACE v

SUMMARY vii

ACKNOWLEDGMENTS xvii

FIGURES xxi

TABLES xxiii

Section
I. INTRODUCTION 1
   Types of Programs 3
   Shortage of Mathematics and Science Teachers 5
   Policy Responses to Problems in the Teacher Labor Market 6
   Research Issues 7
   Organization of This Report 9

II. STUDY FRAMEWORK AND METHODS 11
    Analytical Framework 11
    Methodology 17

III. NONTRADITIONAL TEACHER PREPARATION PROGRAMS 21
     An Overview of Programs 21
     Selected Nontraditional Teacher Preparation Programs 27

IV. NONTRADITIONAL RECRUITS 39
    The Scientific Reserve Pool 39
    Program Recruits 46
    Reasons for Entering Mathematics and Science Teaching 56
    Costs of Participation 57
    Summary 61

V. NEW RECRUITS' PROGRAM EXPERIENCES 64
   Program Attractions 64
   Recruits' Perceptions of Programs 65

	Recruits' Recommendations for Program	
	Improvement	75
	Summary	80
VI.	NEW RECRUITS' TEACHING EXPERIENCE AND FUTURE PLANS	82
	Teaching Status of Graduates	82
	Recruits' Application Decisions	84
	Certification Status and Teaching Position	86
	Recruits' Satisfaction and Adjustment to Teaching	90
	Future Plans	97
	Summary	99
VII.	CONCLUSIONS AND RECOMMENDATIONS	101
	Program Viability	101
	Recruitment Pools Tapped by Nontraditional Programs	103
	Recruits' Preparation Experiences	105

Appendix: DEFINING THE RESERVE POOL FOR MATHEMATICS AND SCIENCE TEACHERS 109

BIBLIOGRAPHY 117

FIGURES

2.1.	The teaching reserve pool: an example	12
4.1.	Trends in beginning salaries for college graduates in selected occupations	41
4.2.	Comparison of salaries of scientist/engineers and teachers	60
4.3.	Comparison of salaries earned by nontraditional recruits in previous jobs	62

TABLES

2.1.	Selected program sample	18
2.2.	Response rates from the survey of recruits in selected programs	20
3.1.	A summary of the study programs	28
4.1.	Degrees and employment in the scientific labor market	40
4.2.	Ratio of expected salaries of college graduates to beginning teachers' salaries	42
4.3.	Selected characteristics of those entering teaching and the total NSF sample: 1982, 1984	44
4.4.	Salaries earned by those who entered teaching and by the total NSF sample	45
4.5.	Number of recruits, by program type	47
4.6.	Demographic profile of recruits, by program type	48
4.7.	Recruits' most recent permanent occupation before program entry, by program type	51
4.8.	Educational background of participants, by program type	52
4.9.	Educational background of graduates, by program type	52
4.10.	Teaching certification status of recruits, by occupation before program entry	53
4.11.	Recruits' prior teaching experience, for those who did not teach in prior year, by occupation before program entry	54
4.12.	Prior teaching assignment of former teachers, by recruit type	55
4.13.	Reasons for interest in mathematics and science teaching, by occupation before program entry	57
4.14.	Recruits' methods of financing program costs, by program type	58
4.15.	Total out-of-pocket tuition costs for program recruits, by program type	59
5.1.	Program features influencing recruits' application decisions, by program type	65
5.2.	Recruits' evaluation of program components, by program type	66
5.3.	Characteristics of recruits' teaching practicum, by program type	69

5.4.	Recruits' evaluation of supervision and assistance received during practicum, by program type	71
5.5.	Recruits' evaluation of program strengths and weaknesses, by program type	72
6.1.	Current teaching status of program graduates	83
6.2.	Recruits' reasons for not teaching after graduating from teacher preparation program	84
6.3.	Importance of criteria in choosing schools to which to apply for teaching positions, by recruit type	84
6.4.	Percentage of recruits who believed they were perceived differently because of the type of program they attended	85
6.5.	Graduates' areas of current certification	87
6.6.	Graduates' current main assignment area	87
6.7.	Percentage of graduates certified in main assignment area	88
6.8.	Selected characteristics of graduates' schools and students, by recruit type	89
6.9.	Graduates' satisfaction with their teaching assignment and desired changes in assignment, by recruit type	91
6.10.	Percentage of graduates satisfied or very satisfied with various aspects of teaching, by recruit type	92
6.11.	Percentage of graduates who considered leaving teaching and their reasons, by recruit type	97
6.12.	Recruits' future plans with respect to teaching, by recruit type	98
A.1.	Selected characteristics of potential reserve pools for mathematics and science teachers, by subgroup	113

I. INTRODUCTION

Over the last several years, there has been considerable and growing concern both about the shortage of qualified precollegiate mathematics and science teachers and the quality of mathematics and science teaching in our nation's schools (Shymansky and Aldridge, 1982; National Commission on Excellence in Education, 1983; National Science Board, 1983; Darling-Hammond, 1984; Carnegie Forum, 1986). The perceived urgency of these problems is heightened by suggestions that the United States will not be able to compete successfully in an increasingly technological world economy without stronger training for many more students in science and mathematics. As a result, many states have enacted legislation increasing the mathematics and science requirements for secondary school graduation. This, in turn, has increased current and future demand for teachers in these fields, thereby exacerbating shortages.

In recognition of the increased demand for teachers and the perceived inability of the traditional teacher pipeline to meet this need, a number of policy initiatives have been implemented. One of these initiatives is the development of programs to train nontraditional recruits for mathematics and science teaching. These programs seek to expand the pool of entrants into mathematics and science teaching by reducing the costs of entry while maintaining the requirement that these recruits be certified to teach their subjects. This report describes a number of these special teacher preparation programs and the "nontraditional" recruits they prepare for elementary or secondary mathematics or science teaching.

Several unique features of the programs and their students set them apart from the usual candidates entering typical teacher preparation programs. We define traditional recruits as "college-age" students who complete undergraduate teacher education courses that satisfy the standard requirements for teacher certification. The programs we have studied differ from this definition in several ways:

- All of them operate outside the context of undergraduate teacher preparation programs: Some are graduate-level programs leading to a master's degree, and others are programs of graduate coursework leading to a teaching certificate but not an additional degree.

- All explicitly recruit "nontraditional" candidates: individuals who have already completed their undergraduate training in a field other than mathematics or science education. These include recent graduates from college with degrees in mathematics or science; individuals who have entered other careers and who are retiring or who want to make a midcareer switch to teaching; individuals who left college some time ago but are not currently in the labor market, such as homemakers; and teachers who initially prepared to teach in areas other than mathematics or science.
- All of the programs provide the coursework necessary for candidates to receive a state teaching certificate. Some of the programs offer the normal series of courses leading to a standard certificate just as "traditional" programs do. Others provide more limited coursework that satisfies the requirements for "alternative certification" (or an "alternate route" to certification) in states that have created such a designation.[1] Still others that train teachers already certified in another field provide courses needed for the added "endorsement" required to teach mathematics or science.

Although we define these programs and recruits as "nontraditional," it should be noted that both the character of "traditional" teacher education programs and the characteristics of their students have begun to change over the last few years. First, while the "college-age" (18–22 year old) cohort of students has begun to decline, undergraduate colleges are enrolling a greater number of older students. Some of these are enrolling in undergraduate teacher preparation programs, thus changing the profile of "traditional" teacher education recruits. One recent study of teacher preparation program entrants in New Jersey,

---

[1]There is much confusion of terms regarding nonstandard requirements for certification recently created by a number of states. Within the last decade, at least 23 states have altered their certification requirements to allow candidates with bachelor's degrees to become certified without taking the full set of education courses normally required for standard certification. Candidates normally take a brief orientation seminar before entering the classroom as full-time teachers and complete additional courses while they are teaching. In some cases, these candidates receive a specially designated "alternative" certificate. In others, the state awards a "full" certificate acquired through an "alternate route"; that is, although the candidate has not completed the normal sequence of courses and credit hours prescribed by the usual state requirements, the candidate taking the special courses earns the same certificate as a candidate in a traditional preparation program. In either case, the certification differs from emergency or limited certification in that candidates have completed a state-approved course of study and are licensed to teach without further restrictions or requirements for additional study. For simplicity's sake, we refer to either "alternate routes" or "alternate certificates" as "alternative certification."

for example, found that the average age of students in traditional programs was 26 years, somewhat older than what is generally considered the typical college age (Natriello et al., 1988).

In addition, graduate-level programs of teacher preparation, although they are not the norm, have long existed and may be increasing in size and number. In 1983, only 3 percent of the estimated 1,287 colleges and universities with teacher preparation programs admitted students for teacher training at the graduate level. Since these were the larger universities, though, they represented a larger proportion of all teacher education students (Feistritzer, 1984). In 1983–84, about 6 percent of recent graduates who had newly qualified to teach received their preparation in master's degree programs.[2] These proportions may begin to grow, since a number of Master of Arts in Teaching (MAT) programs, many of them established during the teacher shortages of the 1960s, have been revived or have recently increased their enrollments (Coley and Thorpe, 1985a), and several states are considering moving all teacher education to the graduate level (Darling-Hammond and Berry, 1988). It may be that in the future, "nontraditional" graduate-level programs like those described in this study will appear less atypical of the modal teacher education program than they do now.

TYPES OF PROGRAMS

Although special teacher preparation and recruitment programs are not limited to those aimed at prospective mathematics and science teachers, they are the most numerous, since shortages in these fields have been the most longstanding and severe. These nontraditional teacher preparation programs can be usefully categorized as follows:

- *Nontraditional recruitment programs* are designed to provide potential teachers from nontraditional pools with the coursework and other requirements for full certification in mathematics or science. These programs do not require changes in state policies regarding teacher training or licensure.
- *Alternative certification programs* are designed to increase the potential supply of teachers by preparing them to meet revised state certification requirements for entering teaching.
- *Retraining programs* are designed to help teachers already trained in other fields to obtain endorsement or certification in mathematics or science.

---

[2]Center for Education Statistics, Recent College Graduates Survey, 1985, unpublished tabulations.

Programs within each of these categories vary considerably in terms of their specific requirements and training components. Nonetheless, certain broad generalizations are possible, as a program's recruitment pool largely determines its design and implementation. Programs focus on the particular kinds of coursework needed for certification by different kinds of recruits. Thus, for example, programs aimed at previous mathematics and science majors usually consist largely of teaching methods courses and internships, whereas retraining programs designed for current teachers in other fields consist almost exclusively of mathematics or science coursework.

Alternative certification programs generally require fewer hours of formal education coursework (generally the equivalent of six to 12 semester hours) but more hours of supervised field experience (often two to four months of teaching under supervision), acquired on the job and with pay. These requirements contrast rather sharply with those of a typical undergraduate preparation program. For example, a typical elementary school teaching candidate must take 36 semester hours of professional studies and 17 semester hours of supervised clinical experience. A secondary candidate is required to take, on average, 25 hours of professional education courses and 15 hours of supervised clinical experience (Feistritzer, 1984).

Another distinguishing feature of many nontraditional preparation programs is that they often offer coursework in a form that is especially designed to build on the previous training of recruits and in a program structure that fits the recruits' practical as well as academic needs. So, course content, program and course schedules, and other supports may be tailored more closely to the particular circumstances of recruit groups. These efficiencies help to lower recruits' transaction and opportunity costs for preparation, making entry into teaching easier.

Although some descriptive data about a few such nontraditional programs have been collected (Coley and Thorpe, 1985b; Adelman, 1986; Fox, 1986), information about their outcomes in terms of teacher training, entry, and retention has not yet been available. This report, along with its two companion papers (Carey, Mittman, and Darling-Hammond, 1988; Hudson et al., 1988), attempts to fill some of these information gaps by providing a detailed look at a sizable number of participants and graduates of a sample of very diverse programs, and at these recruits' motivations and perceptions, program and labor market experiences, and future plans. The data presented here are the first step in providing answers to questions regarding the viability of such nontraditional programs as a partial solution to the immediate teacher supply problem, and as a more long-run means of attracting knowledgeable individuals into teaching.

Nontraditional teacher preparation programs focus on one aspect of the complex set of forces that govern the teacher labor market: the apparent inability of traditional undergraduate teacher education programs to secure enough candidates to meet the demand for mathematics and science teachers. Other factors affecting teacher supply—competition for talent from other labor markets, salary differentials, the relative attractiveness of teaching as an occupation as compared to others—are not directly addressed by these initiatives, although they will undoubtedly have an influence on the success of these programs. To place this policy initiative in an overall context, we begin by briefly delineating the dimensions of the current teacher shortage in mathematics and science and the set of recent policy responses formulated to address this problem.

SHORTAGE OF MATHEMATICS AND SCIENCE TEACHERS

Many recent studies have pointed to an ongoing shortage of qualified science and mathematics teachers. These differ widely in their characterization of the nature and severity of the shortage because of a disparity in definitions, methods, and interpretation. However, there appears to be substantial and convincing evidence that (a) shortages of mathematics and science teachers exist in most states, (b) there is significant out-of-field hiring and assignment of uncertified teachers to teach classes in mathematics and science, and (c) the number of newly trained entrants to these fields has been declining (ASCUS, 1986; Howe and Gerlovich, 1982; NCES, 1983; Carroll, 1985; Shymansky and Aldridge, 1982; Plisko, 1983; Rumberger, 1985; Darling-Hammond and Hudson, 1987; Capper, 1987).

At least two-thirds of the states have reported shortages of mathematics and science (especially physics and chemistry) teachers throughout most of the last decade (ASCUS, 1986; Howe and Gerlovich, 1982). Several indicators suggest that about half of the new mathematics and science teachers hired in recent years are not certified to teach in their assigned fields (Shymansky and Aldridge, 1982; NCES, 1983). Meanwhile, the number of students graduated with bachelor's degrees in mathematics education declined by over 70 percent (from 2,217 to 672) between 1971 and 1982, and the number of degrees granted in science education dropped from 891 to 597. Even with a more recent slight upsurge in response to rising demand, the 1,477 mathematics and science education degrees granted in 1983–84 represent less than one new math *or* science teacher for every 10 school districts in the United States.³ This compares to an estimated demand for mathematics and science teachers over the next several years of about 20,000 annually (Carey, Mittman, and Darling-Hammond, 1988).

Even with other sources of supply (e.g., graduates with degrees in other fields who have minored in education to receive certification, or former teachers returning to the classroom), it is clear that a substantial gap exists between the available supply of mathematics and science teachers and the demand. At the same time, the increasing demand for scientific and technically trained persons in many sectors of the American economy means that teaching must increasingly compete with other occupations for the dwindling supply of college graduates trained in mathematics and the sciences.⁴ As competition for these graduates grows more intense, the already substantial wage differentials between teaching and other occupations are likely to grow even larger, making recruitment of mathematics and science teachers still more difficult.

POLICY RESPONSES TO PROBLEMS IN THE TEACHER LABOR MARKET

The recognition of problems facing the teaching profession has led to a series of policy initiatives and changes. These have sought simultaneously to raise standards—by reforming teacher preparation and licensing—and to recruit more entrants—by improving salaries and offering scholarships and loans for training. Over the past several years, virtually every state has acted to improve teacher pay, producing a 40 percent nominal dollar increase in average salaries nationwide since the start of the decade. At the same time, 46 states have imposed test requirements for entering teaching and most have prescribed teacher education requirements with greater specificity (Darling-Hammond and Berry, 1988). The simultaneous impetus to raise standards and salaries has occurred throughout the twentieth century when teacher shortages have become acute (Sedlak and Schlossman, 1986).

For every one of these moves to tighten certification requirements, however, loopholes allowing candidates to skirt the requirements have

---

³Although this number understates the supply of new teachers by perhaps as much as half, by excluding those who become certified without a major in education, it simultaneously overstates supply by not accounting for the fact that about half of these graduates do not enter teaching, or at least do not do so immediately (Darling-Hammond and Hudson, 1987). Even the most generous adjustments would place the recent supply of newly trained mathematics and science teachers (from traditional programs) at not more than 3,000 annually.

⁴The number of bachelor's and master's degrees awarded in mathematics and life sciences fell by about 25 percent between 1975 and 1985, offset by only slight increases in the numbers of physical science degrees (CES, 1987a).

also been created and expanded. Virtually all states allow candidates to be hired on emergency certificates without any specific training or experience. (More than 20 of the 46 states that grant such certificates allow them to be issued to candidates without even a bachelor's degree (Feistritzer, 1984).) This practice has long been used to fill vacancies during times of recurrent shortages in teaching. During the latest shortages, a number of states have increased their use of emergency certification, and some have created new categories of such irregular licenses. Ironically, states with the greatest number of ostensible screens to entry into teacher preparation and to teaching are those that issue the greatest number of emergency certificates to candidates who have avoided teacher preparation altogether (Scannell, 1988). Although this strategy ensures that classrooms will be filled, it does not attempt to guarantee that those allowed to teach are adequately, or even minimally, prepared to do so competently.

What sets nontraditional teacher preparation programs apart from previous attempts to resolve teacher shortages is that they seek to find a compromise between competing demands for quality and for quantity. That is, they attempt to find, recruit, and prepare for teaching careers individuals who were not attracted to traditional undergraduate preparation programs, while maintaining the requirement that these recruits be certified to teach their subjects. Whereas earlier responses to teacher shortages either relied on efforts to attract recruits to traditional teacher education programs, or skirted these programs entirely by allowing the hiring of uncertified entrants, nontraditional initiatives seek to modify the characteristics of traditional teacher education programs that may serve to limit the numbers and quality of potential teachers they attract.

Individuals who were not ready or willing to commit to teaching during their sophomore year of college—because of differing career goals, financial circumstances, unfavorable labor market conditions for teaching, or other factors at that time—may be unwilling to re-enroll in undergraduate programs to acquire a teaching certificate later. The chance to acquire teaching credentials in a graduate-level program may provide an inducement to entry, particularly when coursework is tailored, credit is given for prior training, and financial aid is available.

RESEARCH ISSUES

The objectives of our research on nontraditional teacher preparation programs are twofold. The first is to identify and characterize programs aimed at preparing mathematics and science teachers, placing particular emphasis on recruitment, program requirements, and conditions that seem important in ensuring their viability. These results are described in a companion note (Carey, Mittman, and Darling-Hammond, 1988) and are summarized in this report, along with a more in-depth examination of a small sample of programs.[5] The second objective of our research is to characterize program recruits, in terms of their backgrounds, motivations, entry into teaching, and future plans. From the vantage point of 481 recruits from nine different programs, we also examine their experiences in teacher training, in the teacher labor market, and in the classroom.

Programs

Previous research (cf. Adelman, 1986; Fox 1986) suggests that some nontraditional teacher preparation programs have been able to attract participants with strong mathematics and science backgrounds, yet the numbers trained in these programs tend to be small, and the longer-range viability of the programs is uncertain. This study addresses the issue of program survival and success at tapping pools of nontraditional recruits, and examines how environmental and policy differences—in certification standards, local labor market conditions, and district personnel practices—may influence program continuation and design. We examined these questions in a survey of 64 programs (reported in Carey, Mittman, and Darling-Hammond, 1988) and a detailed study of nine of these programs, reported here. The specific questions we addressed include:

- What are the different types of nontraditional teacher preparation programs? How do they compare in their content and requirements? What is the current enrollment of these programs?
- How successful have nontraditional programs been in attracting different components of the teacher reserve pool?
- What are the costs of participation in these alternative programs, both in terms of tuition and time? How do they reduce the opportunity costs of training for teaching?
- Are there differences in the survival rates of programs? If so, to what can these be attributed? Have there been recent changes in the programs' design or the target groups they enroll? If so, what sorts of changes have occurred and why? More generally, is it possible to identify factors that influence the viability of a program?

---

[5]The nine selected programs are described in detail in the second companion note (Hudson et al., 1988), and are also summarized here.

Program Participants and Graduates

This report seeks to answer a number of questions about individuals who enter nontraditional teacher training programs, characterizing their backgrounds, motivations, entry, and continuation in teaching:

- What are the educational and occupational backgrounds of recruits? From which parts of the teacher reserve pool are recruits being drawn?
- What attracted these recruits to mathematics and science teaching? Do reasons for entering teaching differ by type of recruit?
- What features of particular programs are important in the decision to enroll? What are the costs to recruits of participation? What sources of financing are available?
- What is the current status of those who have graduated from the programs? Are they currently teaching? What factors were important in their decision to apply to particular school districts?
- How do recruits' actual teaching experiences differ from their expectations? How satisfied are they with different aspects of their teaching job and working conditions?
- What are their future plans? Do they plan to make teaching their career? Do these intentions differ across different types of recruits?
- For those who chose not to teach after graduation, or for those who left teaching, what were the primary reasons for these decisions?

ORGANIZATION OF THIS REPORT

This study surveys the participants and graduates of a selected sample of nontraditional teacher preparation programs. The next section describes the analytical framework and methodology for the study. The third section describes the range of nontraditional programs we studied and the features associated with different types of programs. Section IV assesses potentially productive reserve pools for mathematics and science teaching. We then profile program recruits, both participants and graduates, in terms of demographic and economic variables as well as their reasons for entering mathematics and science teaching. In Section V, we describe their program experiences and how they evaluate the different components of their teacher training. Responses to open-ended questions about what they regard as particular strengths and weaknesses of their program as well as recommendations for making the training more effective are also summarized. Section VI evaluates the proportions of recruits who actually enter teaching and the number who plan to make it a career; it also describes recruits' teaching experiences and future plans. Section VII presents our conclusions and recommendations.

II. STUDY FRAMEWORK AND METHODS

Nontraditional preparation programs aim to entice individuals into mathematics or science teaching who might not otherwise be willing or able to enter. They strive to reduce the various costs of entry—financial costs, transaction costs, and opportunity costs—so that individuals in other occupations or with different training will find it worth their while to prepare for teaching jobs. The various means by which programs target recruits suggest hypotheses about which segments of the labor market can be most readily persuaded and what inducements will be persuasive. In our research, we examine programs and their recruits from the vantage point of incentives offered and costs incurred. Below we describe the framework for our inquiry and the study methods.

ANALYTICAL FRAMEWORK

Potential recruits to teaching can be thought of as constituting a reserve pool. This reserve pool can be divided into components, with different components representing the probabilities of entering teaching that are associated with individuals' varying characteristics. Figure 2.1 depicts these groups of potential recruits in a series of concentric rings: As the rings move out from the center, the likelihood of entry to teaching decreases. Thus, in this illustration, former teachers on leave are more apt to enter teaching than college graduates who did not prepare to teach, since the ease of entry, as well as "tastes" for teaching, would be greater for the former group than the latter. The rings may be further divided into sectors, by age, field, sex, or current activities (e.g., part-time or full-time job, homemaking, education). These sectors also suggest differing probabilities for entry, based on the range of alternative opportunities and the level of opportunity costs facing individuals who might consider leaving their current pursuits to enter teaching. Thus, the conditions under which members of different groups would enter teaching vary.

The examples depicted in Fig. 2.1 are hypothetical, because we do not currently have much information on the propensities of individuals of different types to seek jobs in teaching—or to consider doing so under varying circumstances. Teacher recruitment initiatives, such as the programs studied here, are based on educated guesses about which sectors of the reserve pool may be most productively tapped, and about

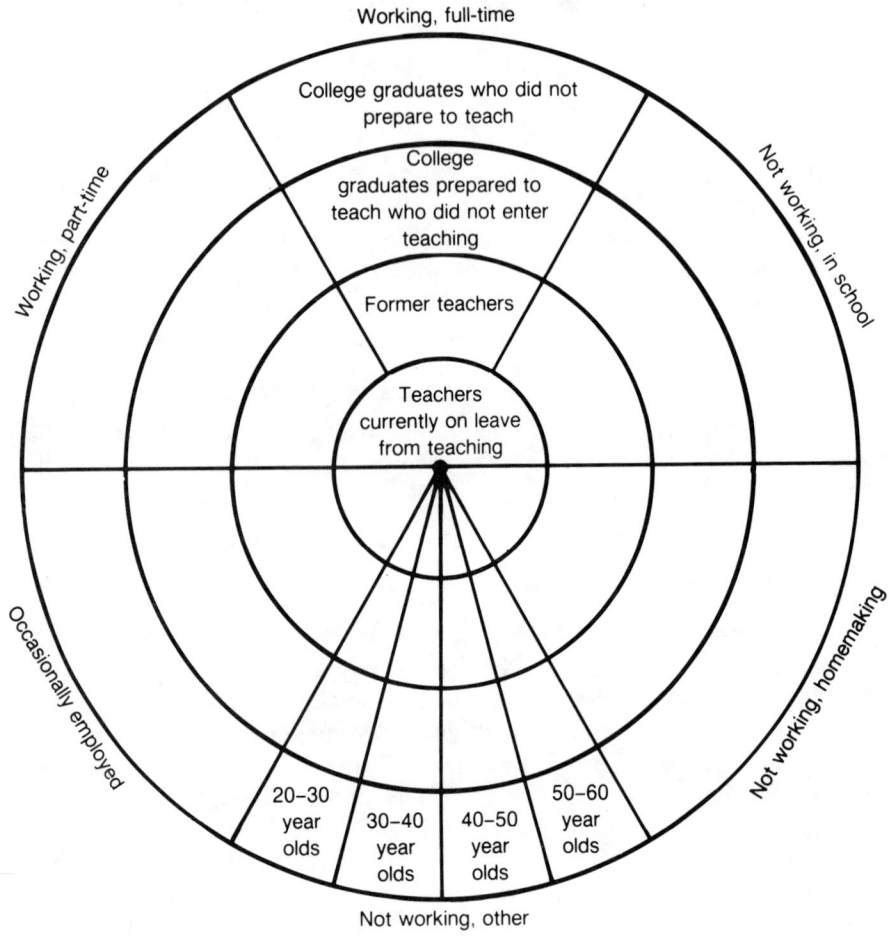

Fig. 2.1—The teaching reserve pool: an example

what recruitment and training strategies are most likely to be successful. This study should help to identify the groups of individuals more likely to teach under varying conditions, thus suggesting policy approaches likely to increase the supply of teachers.

Clearly, nontraditional programs are not designed for and are not likely to appeal to all individuals. Economic theories of occupational choice offer some useful insights in characterizing potentially productive "reserve pools" for mathematics and science teaching. Such theories help explain who might be attracted to teaching by the inducements offered by nontraditional programs in terms of lower opportunity costs of training; they do not, however, address the more complex, and ultimately, perhaps, the more crucial question of *quality*, i.e., whether these recruits into teaching make "good" teachers, by some standard or definition.

Occupational Decisionmaking

The fundamental tenet of the human capital theory of occupational choice is that individuals assess net monetary and nonmonetary benefits from different occupations and, using this assessment, make systematic decisions throughout their careers to enter, stay in, or leave an occupation. Monetary benefits include the stream of likely income resulting from entry into a given profession, likely promotion opportunities, and the value of benefits: health and life insurance, retirement pensions, and so forth. Benefits also include aspects of job security, i.e., the likelihood of steady employment, particularly during periods of generally high unemployment in other sectors. The nonmonetary benefits of an occupation include working conditions, support of coworkers, compatibility of hours and schedules with family and leisure needs, availability of required materials and equipment, and in the teaching profession, such factors as the satisfaction derived from teaching young people.

In addition to occupational income and benefits, an important aspect of any occupational decision is the cost of preparing for an occupation or the costs of retraining for another one. These costs must include the costs of schooling (tuition and living expenses while in school) and forgone earnings during schooling and training. Forgone earnings usually form a significant part of the total costs. The nature of the training one undergoes to enter an occupation and the ease with which these skills can be transferred to another occupation are also likely to play a role in the decisionmaking process.

In simple terms, this theory states that individuals will make decisions to enter or to remain in a particular occupation based on their evaluation of both the costs of training and the benefits from this particular occupation versus other alternatives. Any factor that changes either the costs or the benefits of entering a given occupation or an alternative occupation can cause the individual to reassess the original occupational decision (Grissmer and Kirby, 1987). Nontraditional teacher preparation programs attempt to do precisely that by lowering the costs of entering the teaching profession—both financial, out-of-pocket schooling costs and the opportunity costs presented by forgone earnings during training. For individuals already interested in teaching, this decrease in costs may well tip the balance in favor of entering teaching.

The second part of the theory of occupational choice focuses on the role of uncertainty and incomplete information in determining occupational choices, and on subsequent reversals of these decisions. Like the original occupational choice decision, the process of finding and accepting a job is usually conducted in an environment of uncertainty. The worker and the employer both have incomplete knowledge of each other and of other employment alternatives. However, searching for better or more complete information brings with it attendant costs that must be weighed against the benefits provided by the increased information. For example, continuing to search means sacrificing the income and benefits that could otherwise be earned in the chosen occupation.

Changes in occupation can, from an economic point of view, be explained as resulting from either (a) a reevaluation, based on newly acquired information, of the costs and benefits of the current job versus alternatives, (b) a change in the alternative opportunities available since the original job decision was made, or (c) some combination of both sets of factors. Alternatives may become more attractive either because the chosen job turns out to be less satisfying than hoped, or because the alternative job takes on more appealing features (e.g., wages go up, working conditions improve, access becomes easier or less costly, information about the job and how to attain it becomes more available).

Search models (e.g., Lippman and McCall, 1979) depict the worker as selecting a job without completely searching all alternative jobs because of high search costs. New information on alternatives after job acceptance leads to a reappraisal of the job match. This will be particularly true for individuals with training and skills that are more easily transferred to other occupations or that are particularly valued. Since, for example, degrees in mathematics and sciences are easily transferred to other sectors and are usually better rewarded (at least monetarily) in sectors other than teaching, one could expect mathematics and science teachers to leave teaching more quickly for other jobs. This appears to be the case for highly salable teaching fields like chemistry and physics (Murnane and Olsen, 1988). Conversely, improvements in teaching jobs could lure scientists from their laboratories if more favorable information about such jobs becomes available.

Other models, based on Nelson's (1970) approach to consumer behavior in product markets, assume the existence of unobservable characteristics of a job that can be learned only through experience on the job. As Pencavel (1972) states, "the taking on of a job for a trial period may be the optimum method for an individual to discover whether that employment suits him." In these experience models, then, jobs have two kinds of attributes: inspection (or search) characteristics, which can be observed either directly or without actually experiencing the job, and specific (or experience) characteristics, which become evident only through actual on-the-job experience. Of course, the trial-and-judgment method of occupational decisionmaking becomes increasingly less sensible as the costs of preparation for entry increase. People who have invested heavily in training for their current jobs are less likely to leave for other occupations requiring dramatically different training.

Changes in occupations can then be viewed as the consequence of rational decisionmaking in which belated information regarding the various experience attributes of the current job is received and reevaluated. The theory posits that the greater the amount of experience (versus inspection) attributes present in a given job, the more likely are attrition and changes in occupation. This logic could be used as an explanation for high attrition rates in teaching where, it is often thought, on-the-job experience is needed to ascertain one's suitability or "taste" for the occupation. The theory, though, applies as well to other fields where the nature of the work cannot be well-imagined in advance of entry.

The Decision to Teach

In the context of nontraditional teacher preparation programs, theories of occupational decisionmaking suggest how it is that individuals reevaluate their own and related occupations. From an economic point of view, nontraditional recruits are likely to be individuals for whom opportunity costs, in terms of forgone earnings or value of forgone time, are low. These individuals are likely to be more sensitive to the inducements offered by nontraditional teacher preparation programs in terms of reduced time and financial costs for training. In this sense, the immediate reserve pool may encompass retirees, homemakers, the unemployed and others not in the labor force, as well as recent bachelor's degree recipients who have not acquired a large degree of specific human capital in a particular occupation. These latter are able to switch careers with greater ease than say, individuals with substantial investments in particular professions or firms, precisely because their opportunity costs are low. As will be seen, nontraditional programs attempt to minimize the costs of participation, in terms of both time and forgone earnings. We expect that features such as the duration of the program, availability of student aid, compatibility with current work schedules, and the ability to earn a salary relatively quickly are likely to be important in the decision to enroll.

A number of other factors are also undoubtedly important in the occupational decisionmaking process, including interests and tastes, life cycle stages, and other life circumstances. For example, younger individuals are likely to be more mobile, given that they probably have a lower investment in a particular location than do older individuals with homes and families. They also presumably face lower opportunity costs of retraining for or transferring to other occupations, should they decide to do so. Older recruits probably face a more limited set of alternatives and considerably higher opportunity costs of transferring to another occupation.

General economic conditions are also likely to be important in determining whether individuals will consider changing occupations. If general unemployment increases, for example, job opportunities become more circumscribed and the relative opportunity costs of training for teaching could decline for individuals who are unemployed. High unemployment usually results in lower average wages and salaries in most occupations; this might reduce the disparity between teacher salaries and salaries in other occupations. The expanding teacher shortage in mathematics and science could be an incentive for individuals (including, for example, teachers not currently certified in mathematics or science) facing layoffs or uncertain employment and considering alternative occupations.[1]

The belated information and uncertainty aspect of the theory has particularly important implications for nontraditional programs. There are some grounds for believing that teaching is an occupation with a high number of "experience" characteristics; that is, one cannot really know what the job entails without trying it. Therefore, teaching has very high initial rates of attrition (Grissmer and Kirby, 1987). New teachers, regardless of whether or not they come from traditional teacher preparation programs, often find themselves singularly unprepared for the multiple demands of teaching. Thus, the degree to which nontraditional programs provide recruits with both experiential information about teaching and adequate training to deal with the job's

---

[1]This is, of course, likely to prove true only provided (1) teaching is not also subject to increasing unemployment, and (2) the related mathematics/science fields are the ones hard hit by unemployment, thereby inducing individuals to transfer their subject-matter knowledge to teaching.

demands will greatly influence recruits' retention in teaching. This will also influence program success over the long run, as both potential recruits and potential employers will want to be reassured that recruits' initial job experiences do not lead to high rates of attrition. The basic inference of the theory is simple: The greater the degree of information available to both the new recruit and the potential employer, the more likely we are to recruit and retain these new teachers. This is important, given that the ultimate success of these programs as policy tools must be measured not just in terms of enrollees, but in terms of entry and retention rates.

This economic theory of occupational choice forms the backdrop of our examination of those nontraditional teacher preparation programs designed to attract and recruit a wider pool of individuals to mathematics and science teaching. We turn now to a discussion of our research methods.

METHODOLOGY

Analysis of Nontraditional Programs

To evaluate the success of these new programs at recruiting teachers from different potential reserve pools, we conducted an extensive survey of nontraditional teacher preparation programs across the country in the spring and summer of 1987. Sixty-four programs were identified as meeting the criteria established for the study, i.e., these were programs whose emphasis was on attracting nontraditional recruits and preparing them to be certified as mathematics or science teachers. As mentioned above, these programs include alternative certification programs, other nontraditional recruitment programs, and retraining programs for current teachers.

Information about the programs was obtained both through telephone calls to program directors and through mail surveys. Data collected included basic program descriptions and history, reasons for initiation, sponsors, and targeted recruitment pools. The results of this survey are summarized in the next section; detailed findings are given in Carey, Mittman, and Darling-Hammond (1988).

From the total group of 64 programs, we selected a sample of nine for an in-depth study of participants and graduates. We selected programs from each of the three program categories: alternative certification, nontraditional recruit, and retraining programs. Our criteria for program selection were:

- The program must enroll an average of at least 10 participants. It must have existed for at least one full year at the time the study began; and
- The program must have an accurate list of names and addresses of participants and graduates available.

The final sample consisted of the nine programs listed in Table 2.1. Nontraditional recruitment programs are disaggregated into two groups by their primary target pools—those aimed at midcareer recruits and those aimed at recent bachelor's degree recipients[2]—since differences in the motivations, perceptions, and experiences of different groups are an important focus of our research. Detailed program descriptions, based on on-site interviews with program heads, are provided in the companion note by Hudson et al. (1988).[3] Summary descriptions of these programs are also given in the next section.

Table 2.1

SELECTED PROGRAM SAMPLE

Midcareer programs
  George Mason University: Alternative Science Teacher Program ("Switcher" Program)
  George Washington University: Mathematics and Science Master of Arts in Education Program (MSMA)

Recent B.A. programs
  University of North Carolina at Chapel Hill: Lyndhurst Fellowship Program
  University of Massachusetts at Amherst: Math/English/Science/Technology Education Program (MESTEP)
  West Virginia University: Post-B.A. Teacher Certification Program

Alternative certification programs
  Houston Independent School District Alternative Certification Program (ACP)
  Winthrop College: South Carolina Critical Needs Certification Program (CNCP)

Retraining programs
  Los Angeles County Mathematics and Science Teacher Retraining Program (MSTRP)
  Texas Woman's University: THA-MASTER Program and the Elementary/Secondary Mathematics Teacher Preparation Program

---

[2]These latter include both B.A. and B.S. degree recipients. However, most programs refer to themselves as "recent B.A." programs. We adopt this convention in this report.

[3]In total, 10 programs were originally selected as our sample. Unfortunately, we obtained an extremely low response rate from the survey respondents of one of the programs, because of the inaccuracy of the participant lists. Sixty percent of this sample was invalid because of ineligible respondents and undeliverable mailings. This program was, therefore, deleted from our sample for reporting and analysis. However, we have included the description of the program in the companion note.

Analysis of Recruits' Experiences

Questionnaires were mailed to participants and graduates of the selected programs during June-August 1987. Respondents were asked about their:

- Career backgrounds, including prior teaching experience, salary, and education;
- Reasons for considering mathematics and science teaching as a career and reasons for joining the particular program selected;
- Coursework and training received while in the program and its usefulness in preparing them to be teachers;
- Perceptions of overall strengths and weaknesses of the program and recommendations for improvement; and
- Future career plans.

Program graduates were further asked about their:

- Labor market experiences after graduation, including decisions about whether to enter teaching, criteria for application decisions, and hiring experiences as graduates of a nontraditional program;
- If currently teaching, current teaching assignment, desired changes in assignment, and satisfaction with teaching;
- If not teaching, reasons for leaving or not entering teaching; and
- Future plans.

Table 2.2 presents the response rates for the participant and graduate surveys. Our final response rate was 77 percent. Response rates for individual programs ranged from 63 to 97 percent.

Analysis of the Reserve Pool

Finally, to characterize one major reserve pool for mathematics and science teachers whom these programs seek to tap, we analyzed data from the National Science Foundation's Longitudinal Survey of Scientists and Engineers. For samples drawn from the 1970 and 1980 censuses, we examined how many individuals employed in scientific and technical occupations ever switched to elementary or secondary teaching during the course of the surveys (1972-1978 for the first cohort, and 1982-1984 for the second). We also examined the characteristics of these job switchers and their career paths.

In combination, these several surveys and analyses describe the range of nontraditional recruitment programs, their success in tapping

Table 2.2

RESPONSE RATES FROM THE SURVEY OF RECRUITS IN SELECTED PROGRAMS

Program Site	Sample Size (a)	No. of Completes (b)	Unadjusted Response Rate (%)	Number Undelivered or Ineligible (c)	Adjusted Response Rate (%)
George Mason University	24	20	83	0	83
George Washington University	24	20	83	1	87
University of North Carolina	69	57	83	4	88
University of Massachusetts	75	61	81	2	84
West Virginia University	18	11	61	2	69
Houston	47	31	66	1	67
Winthrop College	109	102	94	4	97
Los Angeles County	240	142	59	15	63
Texas Woman's University	61	38	62	8	72
Total	667	482	72	37	77

NOTE: Unadjusted response rate is calculated as (b)/(a). Response rate adjusted for undeliverable surveys and ineligible respondents is calculated as (b)/((a) − (c)).

various pools of recruits, the motivations and experiences of their participants and graduates, and the character of the potential reserve pool for science and mathematics teaching.

III. NONTRADITIONAL TEACHER PREPARATION PROGRAMS

In this section, we review the findings of a national survey of 64 nontraditional teacher preparation programs[1] and of a more in-depth examination of the nine programs selected for our study of nontraditional recruits.[2] These case studies describe the diverse approaches to teacher training provided by these nontraditional routes.

AN OVERVIEW OF PROGRAMS

Although all nontraditional teacher preparation programs share an ultimate objective of recruitment and preparation of mathematics and science teachers (and, in most cases, teachers in other fields), they vary widely in their approach to this goal. We provide below an overview of how these programs compare on a number of important dimensions, such as their recruitment pools, the content of their training, and program costs. We also briefly summarize conclusions relating to programs' ability to recruit entrants and to survive changes in labor markets and funding.

Recruitment Pools

The typical recruitment pools for these nontraditional teacher preparation programs include various combinations of retirees, homemakers, midcareer transfers, and recent baccalaureates in mathematics or science. An examination of existing programs suggests that members of these recruitment pools may differ markedly in their availability or willingness to enter teacher training programs.

First, retirees participate in nontraditional programs in very small numbers. Programs that have focused on retirees have quickly turned to other sources of participants to maintain their enrollments. This evidence leads us to conclude that, at least presently, retirees are a very limited source of potential mathematics and science teachers. Of course, dramatic changes in program design, or in the working

---

[1]The full survey results are reported in Carey, Mittman, and Darling-Hammond (1988).

[2]These nine programs are described in detail in a companion Note (Hudson et al., 1988); abridged descriptions are provided in this section.

conditions of teaching or the financial status of retirees, may alter this conclusion.

Homemakers also participate in nontraditional programs in very small numbers. This is not surprising, given that historically, relatively small numbers of women have majored in mathematical and scientific fields. Even those who once majored in education are unlikely to have degrees in mathematics or science specialties. Furthermore, homemakers are a shrinking pool from which to draw teachers, given the increasing propensity of women to enter the workforce and to stay in their jobs after having children. Though there may be ways to tap this pool of potential workers for mathematics and science teaching, the current prospects are not encouraging.

In contrast, the survey results suggest that the pools for which alternative preparation programs may have the most promise as distinctive, long-lasting contributors are the midcareer changers and new B.A.s, since programs attract members of these groups in large numbers. Midcareer changers and new B.A.s may also be the most sensitive to the financial and opportunity costs posed by traditional teacher education programs. Postgraduate and master's degree level programs may be much more attractive to them.

Program Content

As briefly discussed in the introduction, the specific content of these nontraditional programs also varies considerably. Alternative certification programs (ACPs) provide the pedagogical coursework to supplement supervised on-the-job experience required to obtain a teaching credential under state "alternative certification" or "alternative route" standards. Alternative certification standards generally require fewer hours of formal education coursework than are required under regular certification standards, but typically require more hours of supervised field experience. In this case, though, "field experience" means full-time teaching as an employed teacher rather than a student teaching experience. Participants usually assume full teaching responsibilities after a several-week introductory course in these programs. Required coursework, conducted in the evenings or on weekends, focuses more on education methods than on subject matter courses, since participants are generally screened for their subject matter background before entering the program.

Nontraditional recruitment programs prepare candidates for standard certification; they tend to include more content area courses than ACPs, and they supplement these with more pedagogical coursework. These programs also include a field placement. In most cases, this placement involves student teaching and observation, rather than complete responsibility for a class. They differ from traditional teacher education programs less in the courses of study or standards imposed than in the organization of coursework into master's degree programs more suitable to mature recruits.

Retraining programs aimed at current teachers consist primarily of mathematics or science coursework. Retrainees already have significant field experience and teaching knowledge, but lack the subject-area knowledge required to obtain certification in mathematics or science. Retraining programs generally provide this knowledge through standard college-level mathematics or science courses, although some retraining programs offer separate classes designed specifically for teachers.

The extent of required study varies substantially across programs and program categories, often depending on the type of teaching certificate candidates earn. The number of required course credits, for example, ranges from nine to 12 (or the equivalent of about three to four courses) among many alternative certification programs (and some others that prepare candidates for provisional credentials) to 40 credit hours or more among master's degree programs targeted at midcareer changers or recent B.A.s. Obviously, the depth and breadth of content coverage varies as well among these widely disparate programs.

Program Costs and Time

Programs also vary substantially in their monetary costs to participants, ranging from free of charge to over $10,000. The Harvard University program for midcareer changers was the most expensive program encountered in this survey, yet it had over twice as many applicants as were accepted. Apparently, Harvard's prestige and preference for a moderately sized program (annual enrollment is about 20) contributes to this program's success at attracting sufficient numbers of highly qualified applicants. In contrast, retraining programs and alternative certification programs, which enroll much larger numbers of mathematics and science teaching candidates, have distinctive financial advantages for participants. These programs are often free, and, in the case of many alternative certification programs, may provide a regular teaching job and salary during much of the training.

Programs differ considerably in their time and effort requirements as well. Nontraditional recruitment programs are similar to traditional teacher education programs—they generally involve a full-time commitment to coursework. At the other extreme, most alternative certification programs include a full-time, fully paid teaching internship following a brief summer course (lasting six to eight weeks). The time spent in teaching methods workshops and seminars is relatively minor when compared to the coursework required in standard certification programs.

The length of time required for program completion varies within program category as well as across categories. Retraining programs and district-based alternative certification programs require somewhat more calendar time for completion on average than the other program categories; this is partially because such programs are designed for part-time participation. In both cases, participants typically hold full-time teaching positions in addition to their formal coursework. Thus, the additional time required to complete this coursework is to be expected. The nontraditional recruitment programs and university-based alternative certification programs, on the other hand, usually involve full-time coursework and field placements. Although the total amount of time spent in formal coursework is greater for these programs, full-time attendance allows participants to complete most programs within 12 to 15 months.

Programs also involve costs to sponsors and the institutions that house the training programs. In some cases where recruits pay little or no tuition, all of the costs are borne by third parties. Though we could not collect detailed information on total costs (and many costs are hidden or indirect), the range of third-party costs seems to be from $2,500 to $10,000 per candidate, depending largely on the program's duration and intensity. In addition, tuition charges are substantially offset by scholarships or loans to students, or by paid internships. The funding sources vary widely, including state and local district funds (especially for alternative certification and retraining programs), federal government and foundation grants, and contributions from industry. In some cases, university subsidies to programs are substantial as well. This reliance on outside funding has made some programs quite vulnerable to funders' changing priorities.

Contributions to Teacher Supply

The survey located 64 programs that reported enrollments of 2,443 prospective mathematics and science teachers in 1986–87. This does not represent the entire universe of such programs, but we believe it probably includes a sizable majority of existing programs. Considering that over 20,000 new mathematics and science teachers will be needed each year over the next decade,[3] the graduates of these initiatives represent a small but nontrivial fraction of the total needed, perhaps 10 percent or more. Some local and state programs supply as many as 15 to 30 percent of new teacher hires in those jurisdictions for certain subjects. Nevertheless, these programs do not, by themselves, attract enough participants to solve the shortage of qualified mathematics and science teachers. The rate at which programs are created, modified, and ended suggests that the share of teaching positions they help to fill will continue to be unpredictable, at least in the short term.

It is worth noting that of the 64 programs we identified and located for the survey, eight enrolled no students or had been discontinued by the time we completed the survey; several others were unsure whether they would continue operating in the following year. The causes of these difficulties include labor market and funding changes, as well as inability to find and recruit enough target candidates.

Changes in Program Design

The survey also found that many programs change in response to the willingness of particular recruit pools to enter teaching and of local districts to hire new recruits. In particular, some programs that originally focused on retirees have already begun to focus on midcareer changers and new B.A.s as a more ready source of applicants. Other programs that orginally focused exclusively on mathematics and science have opened to applicants in other subject areas. The pressure to attract applicants from a wide range of sources to maintain enrollments, and the pressure to be responsive to quickly changing districts needs, provide incentives to broaden the programs' emphasis. This broadening may result in a loss of identity with particular applicant pools or subject areas.

Many university-based programs also become institutionalized over time as part of the university's regular master's degree or certificate training programs. This can have beneficial effects for both the "traditional" and "nontraditional" programs. The latter are more protected from the vicissitudes of funding and labor market shifts as part of a standard university program. The former often learn from the success of nontraditional programs in packaging training to better meet the scheduling and coursework needs of different target populations. In addition, the successful master's degree programs ultimately encourage

---

[3]This estimate is based on a number of assumptions regarding teacher demand and supply. Should these assumptions be inaccurate, the actual number could vary considerably from our estimate.

the trend for teacher education to become a postgraduate professional training activity.

Factors Affecting Program Viability

Several generalizations about program viability can be made on the basis of program changes noted in the course of the survey. Survey results suggest that a program's success in maintaining enrollments is strongly affected by the state of the local or regional labor market, local demand for mathematics and science teachers, and program funding. Changes in unemployment rates appear to be particularly important in determining interest in the teaching profession, with a lack of opportunities in industry contributing to greater interest in teaching careers (and sometimes to a temporary disappearance of teacher shortages). This has occurred most obviously in the Southwest, when unemployment in the oil industry contributed to a greater pool of mid-career scientists willing to try teaching, at least until industry fortunes changed.

Programs also operate largely at the whims of local teacher demand; when demand in an area drops, programs are likely to cease operation, contract in size, or shift training to areas of greater need. The Houston Alternative Certification Program, for example, dropped mathematics and science training after its first year because the Houston region is now experiencing a much greater need for elementary and bilingual instructors, and has been able to fill many of its mathematics and science vacancies through emergency hiring and a reassignment of elementary teachers to the secondary level. Finally, discontinuation of outside funding has led to the closing of many programs, although in some cases (e.g., University of North Carolina's Lyndhurst Fellowship Program), the sponsoring university incorporated the program, or important program features (such as part-time course scheduling or an advanced pace) into its regular teacher training program offerings.

Generally, programs that have survived and maintained their enrollments have done so by remaining flexible in how they seek funding, in who they recruit, and in how they package their programs. It appears that (a) nontraditional programs may be most successful at attracting new mathematics and science teachers in times of high or rising unemployment in math/science fields outside of education; and (b) programs are more successful when they are planned with a clear understanding of the needs of local school districts, when they are flexible in targeting recruit pools, and when they can be supported by regular teacher education programs.

SELECTED NONTRADITIONAL TEACHER PREPARATION PROGRAMS

Nine nontraditional programs were selected from the larger sample of 64 programs for more detailed examination, including a survey of program participants and graduates. The basic features and structure of these selected programs, as they existed in 1987, are described below and summarized in Table 3.1.

Midcareer Programs

George Mason University: Alternative Science Teacher Program (Switcher Program). The Switcher Program is a full-time, one-semester (16 week) program designed to prepare individuals with prior academic training and work experience in earth science, chemistry, or physics to teach these subjects in grades 8–12. It was begun as a means of alleviating shortages as well as improving the quality of science teachers in the area. It started in January 1986 with eight students; the current number of participants is between 12 and 15 per semester. Admissions criteria include a bachelor's degree in the relevant field, three years of work experience, and at least a 2.75 GPA in the last two undergraduate years of college.

Participants complete nine semester hours of coursework (three courses) in education during the first eight weeks of the program, followed by six semester hours of student teaching during the second eight weeks of the program. This student teaching is done at one of three participating local school districts under the supervision of cooperating teachers and program faculty. Students take on increasing levels of classroom responsibility during the practicum. They graduate from the program with provisional certification and are eligible for full certification after two years of teaching.

The program is funded primarily by a three-year grant (of about $100,000) from the Office of Educational Research and Improvement of the U.S. Department of Education. In addition, the Virginia Department of Education provides a $1,000 forgivable loan to all in-state participants. This covers approximately the full cost of the program for these students.

The program graduated 14 students in the spring and fall of 1986. Of these, nine were teaching in 1987, although program administrators expected most of the others to teach eventually (many were completing the additional coursework they needed for certification, but that was not provided by the program).

Table 3.1
A SUMMARY OF THE STUDY PROGRAMS

Selected Characteristics	Midcareer Programs		Recent B.A. Programs			Alternative Certification Programs			Retraining Programs	
	Switcher Program GMU	MSMA, GWU	Lyndhurst Fellowship Program, UNC	MESTEP, UMass	Post-BA Certification Program, WVU	Houston ACP	CNCP, Winthrop College	L.A. County MSTRP	Texas Woman's University [a]
Outcome	Provisional certification (Va. law); after 2 years of teaching, eligible for full certification	Master of arts in education; certification. Some may choose to meet only certification standards (18–21 credit hours)	Master of arts in teaching and "G" level certificate	Master of arts in education; certification in one subject area	Master of arts in education; certification	Certification	5-year regular certificate	Supplementary authorization (equivalent to full certification)	Elementary/secondary credential to teach mathematics
Funding (a) Cost (tuition)	$1,000	$10,800	$1,300 in state; $7,000 out of state	$4,000	$630 in state; $1,670 out of state	First year free	9 hours of graduate coursework, otherwise none	None	$1,078
(b) Student aid	Va. provides $1,000 forgivable loan to in-state residents	Washington, D.C., pays tuition for those willing to work in D.C.; V.A. benefits for military participants	$7,500 stipend plus paid tuition	$9,000 per semester as teaching or industry interns. Student loans available	Need-based tuition waivers and student loans	$19,000 starting salary	Salaries for both years as a beginning teacher	All current teachers receive regular salaries	1 district paid tuition for its teachers; all current teachers receive regular salaries
Future plans/other comments	Expand enrollment	Add content area refresher courses; add round-table discussions with experienced teachers	Program ends June 1988. School of Education may change its MAT program to make it more similar to the Lyndhurst Program	Obtain more secure funding; improve quality of staff; track graduates	Full-time students only; an "intern" year for new teachers; use of "modules" rather than semester courses	Program now focuses only on elementary certification	More formal training of local school mentors; more field coordinators	No longer funded	THA-MASTER ended Summer '86. The model is being used by TWU to train mathematics teachers in new program

Table 3.1 (continued)

	Midcareer Programs		Recent B.A. Programs			Alternative Certification Programs		Retraining Programs	
Selected Characteristics	Switcher Program GMU	MSMA, GWU	Lyndhurst Fellowship Program, UNC	MESTEP, UMass	Post-B.A. Certification Program, WVU	Houston ACP	CNCP, Winthrop College	L.A. County MSTRP	Texas Woman's University[a]
Recruitment pool	Midcareer transfers, military and other retirees	Midcareer transfers, military and other retirees	Those with B.A. degrees in relevant fields (or equivalent coursework)	Recent B.A. in mathematics, science, or English	Those with B.A. degrees in relevant fields	Those with B.A. degrees and 24 credit hours in mathematics or science	Midcareer transfers, recent B.A.s, retirees with degrees in math, science, library science,	Current teachers	Current teachers
Date program started	Spring 1986	Fall 1985	July 1982	June 1983	1973, changed last 3 years	Fall 1985	May 1985	1982	Summer 1985
No. participants	8-12	5-8	13-18	19-25	15-25	350	15-65	100-170	23
Length of program	16 weeks (full-time)	2 years for M.A. (part-time)	1 year (July 1-June 30) (full-time)	15 months (June-August) (full-time)	15 months (1 academic year and 2 summers) (full-time)	Approximately 12 months	2-3 years	2 years (part-time)	14 months (1 academic year and 1 summer) (part-time)
Preparation for:									
(a) Subjects	Earth science, chemistry, physics	Mathematics, science	Mathematics, science, English, social studies	Mathematics, science, English	Mathematics, science, English, foreign language	Elementary, English, math, science, etc.	Mathematics, science, library science	Mathematics, science	Mathematics
(b) Grades	8-12	7-12	9-12	7-12	K-8, 5-8, 5-12, 9-12	K-12	7-12	7-9	5-9
Program requirements									
(a) Coursework	14 credit hours in education (8 weeks) and student teaching	36 credit hours in education and student teaching	40 credit hours, 16 in education, 18 in content area, 3 in elective and student teaching	51 credit hours in education, teaching internship, and industry internship	51 credit hours in education and student teaching	Training in August (40 hours); 6 one evening a week during school year (100 hours); following summer (30 hours); all in education	2-week institute before teaching; 8 monthly seminars during first year; another 2-week institute in second summer; 3 additional educ. grad. courses	5 courses in mathematics or science	18-24 credit hours in mathematics
(b) Practicum	8 weeks	10 weeks	2 semesters	1 summer + 1 semester. Latter involves full responsibility for classroom	1 semester	1 year of regular teaching with full responsibility for classroom	2 years of regular teaching with full classroom responsibility	None	None

[a] Includes two programs: THA-MASTER (The Hellman Academy for Mathematics and Science Teacher Education Retraining) and the Elementary and Secondary Mathematics Teacher Preparation Program. The latter program was modeled on the former.

George Washington University: Midcareer Mathematics and Science Master of Arts in Education (MSMA) Program. The MSMA Program is an academic program designed for individuals with work experience and undergraduate or graduate training in mathematics or science. It was begun in response to complaints by local school districts of shortages and shrinking applicant pools for mathematics and science teachers and substitutes. Because of its focus on impending retirees from the military or government, it is a part-time program with coursework scheduled around participants' work hours. The program began in the fall of 1985 with eight participants. In 1987, the program had 22 active students. Program applicants must meet GWU graduate school admissions criteria. These include a bachelor's degree, a minimum 2.75 undergraduate GPA, a specialization in one of the relevant fields, and a score ranking above the 50th percentile on the Miller Analogies Test or Graduate Record Exam.

Graduates receive a master of arts in education; alternatively, they may choose to remain only until they meet certification standards. The program typically takes two years to complete part-time or three semesters if enrolled full-time. Degree candidates must complete a total of 36 credit hours, all of which are education or education-related courses; certification-only candidates must complete 18–21 credit hours of coursework. The field component consists of 10 weeks of student teaching under the joint supervision of a master teacher and a university faculty member. Students take on increasing levels of responsibilities for the classroom during the practicum.

The program operates under the University's School of Education and Human Development. The full cost to participants of the degree program is $10,800. The program itself does not offer financial aid but students may be eligible for Veteran's Assistance benefits or for tuition assistance contributed by the District of Columbia Public Schools for teachers willing to work in the city.

As of spring 1987, the program had graduated three students, all of whom were teaching.

Recent B.A. Programs

University of North Carolina at Chapel Hill: The Lyndhurst Fellowship Program. The Lyndhurst Fellowship Program is a fifth-year MAT program for arts and science baccalaureates interested in teaching mathematics, science, English, or social studies. The two main goals of this experimental program are to: (1) attract highly talented people with degrees in subjects other than education into mathematics and science teaching, and (2) experiment with the curriculum to see if the number of education courses could be reduced without jeopardizing students' ability to become effective teachers.

The program began in July 1982 with 22 participants. To increase the stipend level to students and to allow for more innovation with the practicum, the size of the program was subsequently capped at 18 participants. Students are prepared to teach grades 9-12. Program applicants must meet UNC graduate school admissions criteria. For those with no teaching experience, these requirements are a minimum 3.3 GPA and 1150 GRE score. Participants receive a master of arts in teaching and full certification on successful completion of the program. The program takes one full calendar year to complete: two academic semesters and two summer sessions. All students attend full-time. Participants must complete 40 credit hours of coursework, including pedagogy (18 hour minimum) and subject-area courses (16 hour minimum). The course work includes a two-semester practicum under the direction of a mentor teacher and an education faculty member.

The program is part of the graduate school at UNC, within both the School of Education and the College of Arts and Sciences. It is approved by the North Carolina Department of Instruction as an experimental program. The major share of the budget (85 percent) is provided by the Lyndhurst Foundation. All students receive a $7,500 stipend plus paid tuition. The program requires students to teach for three years after graduation, or repay the stipend. About 75 percent have complied with this teaching requirement. The program ended in June 1988, but is continuing as a "fast track" option as part of the university's regular MAT program.

University of Massachusetts at Amherst: Math/English/ Science/Technology Education Program (MESTEP). MESTEP is a 15-month program of academic coursework and internships in teaching and industry, designed for recent baccalaureates with degrees in mathematics, science, or English. The emphasis on *recent* B.A.s is meant to ensure that participants are current in their field. The program was initiated in response to expressed recruitment concerns of local school superintendents. It began in 1983 with 19 participants, and currently accepts between 19 and 25 students annually. Students must be accepted into the UM Graduate School, which requires a minimum 2.75 GPA. Applicants are also interviewed by participating school districts and industry personnel, who must be willing to hire the applicant for an internship position.

Program participants take 51 credits (including internships) to receive a master's degree in education and full certification. Students begin in June at Amherst, with three intensive courses in methods, planning, and assessment. They then transfer to the eastern section of the state, where they student teach in July and August, then have a one-semester teaching internship and a one-semester industry internship during the school year, while taking two evening courses and a monthly seminar. Finally, students take four more courses during the second summer; these courses are geared toward both education and research. The teaching internship is under the supervision of a mentor teacher and university supervisor. Most industry internships involve job training or other education-related work (e.g., training curriculum development). It is anticipated that many graduates will keep working in these industry internships during the summer months; they are designed as an incentive to help retain teachers in the classroom, by providing them with more varied teaching career options.

The program's funding sources have varied, and have included the University's School of Education, grants from the U.S. Department of Education's Fund for the Improvement of Post-Secondary Education, and the Massachusetts Board of Regents. Tuition for the 15-month program is $4,000; however, interns are paid about $9,000 per semester.

As of spring 1987, the program had 57 graduates, 80 percent of whom had gone into teaching. Participants agree to teach for three years when they enter the program; about 70 to 75 percent of the graduates comply.

West Virginia University: Post-B.A. Teacher Certification Program. This program provides a master's degree in education and full certification to individuals with a content-area degree in English, foreign languages, mathematics, or science. The WVU College of Education began the program in 1973 as an "alternative" certification route; the program has changed over time in response to changes in the state's certification requirements (i.e., meeting these requirements has now become less time-consuming). The program was begun for two reasons: (1) There were teacher shortages in West Virginia and (2) WVU had a supply of interested and available teacher candidates in the form of university-affiliated personnel (e.g., faculty spouses, laboratory technicians). In the early years, the program was much smaller than its current size of approximately 25 participants per year.

To be admitted into the program, an individual must meet graduate school entry requirements—a minimum 3.0 GPA, relevant subject-area degree, minimum score of 1000 on the Graduate Record Examination, and a demonstrated interest in teaching. Applicants must also pass the state-mandated Pre-Professional Skills Test. In theory, participants can prepare to teach grades K–8, 5–8, 5–12, or 9–12; most prepare for the latter two grade spans. Students must take 27 hours of education coursework, including one semester of student teaching. Those wishing to obtain a master's degree take an extra nine credit hours in their content area, for a total of 36 credit hours. Courses may also be taken on a full-time or part-time schedule. Participants student teach for one semester, under the joint supervision of the cooperating district teacher and university faculty. After completion of all course requirements, participants must pass the state-mandated content test in the area of specialty before being recommended for certification.

The program is part of WVU's College of Education and has not, thus far, received outside funding. Tuition is $630 per semester for in-state residents and $1,670 for out-of-state students. The overall cost for the 36 credit-hour master's program is approximately $2,500 or $7,000 (depending on state residency).

The program had about 150 graduates by spring of 1987, most in areas outside mathematics and science. Over 80 percent of all graduates take teaching positions.

Alternative Certification Programs

Houston Independent School District Alternative Certification Program (ACP). The Houston ACP was designed to address widespread teacher shortages in the Houston area by attracting well-qualified candidates to teaching. The program allows participants to fulfill state credentialling requirements by (a) serving a one-year internship under the supervision of an experienced, certified teacher and (b) taking classes in teaching methods and classroom management. The program is mainly directed toward working professionals seeking to change careers, homemakers, and retirees. The program began in the fall of 1985 with 350 participants. Candidates for mathematics and science teaching must have a bachelor's degree, and at least 24 credit hours and a 2.5 GPA in the subject they wish to teach. All participants are also required to pass the ETS Pre-Professional Skills Test or the Texas Functional Academic Skills Test, as well as a subject area test (for secondary candidates).

During the first year of the program (the only year in which teachers were trained for mathematics and science), candidates participated in a preassignment orientation in August in which they received 40 clock-hours of training in lesson planning, classroom management, etc., and observed a classroom for a full week (25 hours). Individuals offered a contract by a Houston school then began full-time teaching, under the supervision of a senior teacher. Over the course of the year, participants received about 100 clock-hours of individualized training; these courses (modules), developed by district personnel, focused on classroom management and other pedagogical areas. The only subject matter instruction was at the end of training, as a refresher before taking the EXCET test (required of all Texas teacher candidates). In addition, interns took 30 hours of workshops during the following summer, covering classroom management and teaching methods. The program thus totals 170 hours of coursework, roughly the equivalent of 12 semester-hours, or four college courses.

The program was initiated by the district, with the district providing all funding. Participants are paid at full salary levels for beginning teachers. In the first year, participants were paid to attend the training sessions. During the second year, participants were asked to bear some expenses, and stipends were lowered.

Of the approximately 350 entering the program in the first year, about 200 graduated; of these, 155 received their state credential, and approximately 50 students were still working toward it in 1987. Graduates tend to continue teaching in the schools to which they were initially assigned. In the first year, the program had 45 mathematics and science participants.

There have been considerable changes in the program since the first year. The focus is now on those seeking elementary and bilingual credentials, more pressing shortage areas for the city. Other changes include increases in the credit hours required for program completion (to meet new state requirements), and having local universities provide some of the training.

South Carolina Critical Needs Certification Program (CNCP). In 1984, South Carolina's Educational Improvement Act established the Critical Needs Certification Program to recruit and train teachers in designated shortage areas (currently mathematics, science, and library science). The CNCP is directed toward midcareer transfers, retirees, and recent graduates with bachelor's degrees in the relevant fields. The legislation allows participants to begin teaching provided they have at least a bachelor's degree in a "critical needs" area and have passed the appropriate specialty section of the National Teacher Examination. Participants who meet these requirements receive a letter of eligibility from the State Department of Education; they may then seek employment with a public school district in South Carolina. The CNCP began in July 1985 with 15 participants. There were 142 active participants in 1987.

The program has three distinct phases: The first phase consists of an intense two-week institute focusing on pedagogical principles and methodological skills; this precedes the first year of teaching. Upon successful completion of this coursework, participants are issued a conditional teaching certificate. During the course of the school year, participants (who are now full-time teachers) attend eight monthly seminars, also focusing on teaching methods and classroom management.

Participants must pass three performance assessments (required by the state) and all subtests of the South Carolina Educational Entrance Examination. Participants who meet these program requirements earn three hours of undergraduate credit from Winthrop College.

The second phase includes a second two-week summer institute (following the first year of teaching) and a second year of teaching. Under new regulations (as of July 1987), participants are supervised by an in-school mentor and a college coordinator. Successful completion of phase two enables participants to earn six hours of graduate credit from Winthrop College. In the third phase, candidates must complete three additional graduate courses within three years of the issuance of the conditional certificate. Upon completion, participants receive a regular five-year professional South Carolina teaching certificate for grades 7-12.

The program is administered by Winthrop College; however, both the State Department of Education (SDE) and local districts play significant roles in its implementation. For example, the SDE sets admissions and retention criteria. The only cost to participants is tuition for the nine hours of required additional graduate work; the program budget covers all other participant costs, including books, housing, and food.

Of the 142 participants, 61 entered mathematics teaching and 79 entered science teaching.

Retraining Programs

Los Angeles County Mathematics and Science Teacher Retraining Program (MSTRP). This was a part-time program to retrain current teachers (including those currently teaching mathematics or science on emergency credentials) for endorsement in mathematics and science at grades 7-9; state funding for the program ended in early 1988 (after our surveys had been completed).[4] The two-year program was initiated by the California Department of Education to address the severe shortage of qualified mathematics and science teachers in Los Angeles County. The program began in 1982 with 170 accepted participants. Acceptances were based on whether applicants had good backgrounds in mathematics or science, and whether they lived within commuting distance of a retraining site (i.e., participating university). The program's last cycle had two science retraining sites and four mathematics retraining sites. The program consisted of five

---

[4]The budget for the L.A. County Office of Education's Teacher Education and Computer Center (TECC), under which this program was operated, was eliminated because of state budget cuts.

subject-area courses, one per semester; these were equivalent to regular college courses and generally met twice per week. The program also included a series of "content delivery" seminars developed by the county on such topics as concept, skill, and language development. All participants were full-time teachers in L.A. County, so the program did not include a field placement.

The L.A. County Office of Education subcontracted with five participating colleges to provide coursework. The program was fully funded by the California State Department of Education; participants did not pay for tuition or books.

Approximately 69 participants (about 50 percent of enrollees) completed the first program cycle; another 62 (62 percent of enrollees) graduated from the second cycle. These high rates of attrition occurred primarily because of the combined demands of work, traveling, homework, and family responsibilities. Data are not available on the number of teachers switching to mathematics or science teaching following program completion.

Texas Woman's University: THA-MASTER Program and the Elementary and Secondary Mathematics Teacher Preparation Program.[5] These programs were designed for the retraining of teachers from other disciplines or grade levels to become elementary and secondary school teachers of mathematics. The programs were initiated because of shortages of mathematics teachers in the Denton and Dallas-Ft. Worth areas. The THA-MASTER Program ran for 14 months, from the summer of 1985 to the summer of 1986, at which time the funding cycle (from the Fund for the Improvement of Post-Secondary Education) was completed. At this point, TWU instituted the Elementary and Secondary Program, using Title II funds, to continue the retraining program; both programs are very similar, the major difference being that the THA-MASTER Program enrolled participants as a cohort, whereas participants in the Elementary and Secondary Program go through the program individually. Both programs were also designed to be part-time in the winter and full-time in summer, to accommodate the needs of full-time teachers.

The THA-MASTER Program had 23 participants; the Elementary and Secondary Program began with 40 participants and currently enrolls between 24 and 44 participants annually. Candidates are required to have an undergraduate degree with a minimum 2.75 GPA (needed for admission to the graduate program), teaching experience, and a desire to teach mathematics and to stay in Texas. Participants

---

[5]THA-MASTER stands for The Hellman Academy for Mathematics and Science Teacher Education Retraining.

are required to complete a total of 18 semester credit hours in mathematics to obtain an elementary credential, and 24 credit hours to obtain a secondary mathematics credential. Courses offered include survey of mathematics, applied calculus, and current issues in mathematics education. There is no field experience requirement as these participants have been or are currently teachers.

Both programs were or are affiliated with the university's Department of Mathematics, Computer Science, and Physics. The tuition cost of the current program is approximately $1,100.

Overview of Selected Programs

In addition to these individual program summaries, the interviews with program administrators yielded several key observations and generalizations about these nontraditional teacher preparation programs.

First, as Carey et al. (1988) also noted, the programs that focus on special populations of recruits tend to exist in locations where teacher demand is high and a viable recruitment pool is easily accessible. For example, the two programs targeted toward midcareer transfers and retirees (George Washington University and George Mason University) are both located in the Washington, D.C., area, where teacher shortages exist, and where there are a large number of research, technical, and military industries (see GMU program description in Hudson et al., 1988).

Even given a "high supply" location, these programs usually must actively recruit to obtain a sufficient number of well-qualified participants. Active recruitment is necessary to attract individuals who are typically unaware of the existence of a teacher preparation program that is targeted and designed for them; the newness of these programs, as well as of the concept underlying them, requires extensive advertising and recruitment. Many programs also intensively recruit to maintain stringent entry requirements. For example, the University of Massachusetts, West Virginia University, George Washington University, and the University of North Carolina programs all use graduate school admissions requirements. The George Mason University program also has strict academic *and* work requirements.

Administrators typically felt that their recruitment efforts are very successful, although many mentioned that these efforts use a significant amount of programs' limited financial resources. Some, such as the University of Massachusetts, have been especially successful in recruiting minority candidates to these fields in which they are seriously underrepresented. Over time, some programs have found that they have been able to decrease their recruitment efforts without sacrificing program quality.

Programs targeted toward retirees and midcareer transfers have sometimes found that recruitment is made more difficult by teaching's relatively low salary level; many potential candidates lose interest once they discover how much a teacher actually makes. This appears to be especially problematic in areas where military retirees have a wide range of opportunities in private industry.

Another common program feature that is viewed as a strength by most program administrators is admitting participants in cohort groups and structuring interaction among members of each group. Training participants as a cohort is viewed as a means of maximizing participants' shared knowledge and support, and is usually seen as one of the most important parts of the program's learning process. Administrators at the University of Massachusetts, for example, felt that cohort groups served to "strengthen and energize" participants. Two of the three programs that do not use cohort groups mentioned that they would like to use them, and do in fact have plans to increase interactions among program participants.

Finally, nontraditional recruitment programs tend to exist at institutions where experimentation with teacher training is common and accepted. For example, the two universities sponsoring the midcareer programs also have "regular" fifth-year MAT programs, as does the University of North Carolina. All institutions sponsoring recent B.A. programs are affiliated with, or use models provided by, recent teacher education reform initiatives: West Virginia University and the University of North Carolina are members of the Holmes Group, and the University of Massachusetts is structuring its nontraditional program around the recommendations made by the Carnegie Forum on Education and the Economy (Carnegie Forum, 1986).

IV. NONTRADITIONAL RECRUITS

Programs aimed at recruiting new entrants to mathematics and science teaching are seeking to tap different supply pools for teaching. Aside from current teachers and college students, a major source of potential talent is those individuals working in or trained for scientific fields who might be persuaded to use their backgrounds in a new way. The hope of many nontraditional programs is that the pool of scientific and technically trained personpower will yield a number of recruits who are looking for a change at the beginning, middle, or end of their careers.

In this section, we look at this recruitment pool, the characteristics of its members, and their propensities in recent years to enter elementary and secondary teaching. This examination describes a substantial portion of the reserve pool for mathematics and science teaching that might be tapped by innovative preparation programs.

Following this overview of a major sector of the reserve pool, we present a demographic and economic profile of those individuals who enrolled in the nine nontraditional teacher preparation programs we studied. These data provide basic information on who in fact enters nontraditional programs and on their reasons for wanting to enter mathematics and science teaching.

THE SCIENTIFIC RESERVE POOL

For more than a decade, the labor market demand for scientifically trained workers has been sharply increasing. Between 1976 and 1986, employment in engineering and physical and life sciences nearly doubled, and employment in computer science fields increased almost fourfold. Degree production in computer science and, to a lesser extent, engineering has been largely keeping up with this demand. But in the life sciences, physical sciences, and mathematics, the number of earned degrees awarded by U.S. universities has been stagnant or declining (see Table 4.1.) Furthermore, a growing share of these degrees—ranging from 20 to 40 percent—has been awarded to foreign national students, many of whom return to their countries after graduation (Vetter and Babco, 1987).

These data suggest that many other scientific occupations are likely to be competing with each other and with teaching for a dwindling supply of newly trained entrants, especially in the fields that rely on

Table 4.1

DEGREES AND EMPLOYMENT IN THE SCIENTIFIC LABOR MARKET

	Computer Sciences	Engineering	Life Sciences	Physical Sciences	Mathematics
Earned degrees[a]					
Bachelor's—1975	5,033	39,388	51,741	20,778	18,181
Bachelor's—1985	38,878	77,154	38,445	23,732	15,146
(% change)	(+672%)	(+96%)	(−26%)	(+14%)	(−17%)
Master's—1975	2,299	15,127	6,550	5,807	4,327
Master's—1985	7,101	20,926	5,059	5,796	2,882
(% change)	(+209%)	(+38%)	(−23%)	(−0%)	(−33%)
Doctorate—1975	213	3,106	3,384	3,626	975
Doctorate—1985	248	3,221	3,432	3,403	699
(% change)	(+16%)	(+4%)	(+1%)	(−6%)	(−28%)
Employment[b]					
1976	116,000	1,278,300	198,200	154,900	43,800
1986	437,200	2,243,500	340,500	264,900	103,900
(% change)	(+277%)	(+76%)	(+72%)	(+71%)	(+137%)

[a]SOURCE: CES (1987a), pp. 190–192.
[b]SOURCE: National Science Board (1987), p. 218.

training similar to the most prevalent secondary school teaching specialties—mathematics, biology, chemistry, and physics. Given the wage disparities that exist between teaching and other scientific and technical occupations (Fig. 4.1), it seems clear that the task of recruiting teachers from this pool is not a trivial undertaking. Though beginning wage differentials for recent college graduates entering teaching compared with other fields have decreased somewhat since 1982, the gap is still larger than 30 percent between teaching and such fields as engineering, chemistry, mathematics, and computer science. (See Table 4.2.)

Still, many factors other than income potential motivate initial career choices and later changes. We examined the extent to which individuals trained for or employed in scientific occupations move into teaching by analyzing data from the National Science Foundation's (NSF) longitudinal registry of scientists and engineers. This registry provides information on personal characteristics, education, and current and previous employment for a sample of those identified as

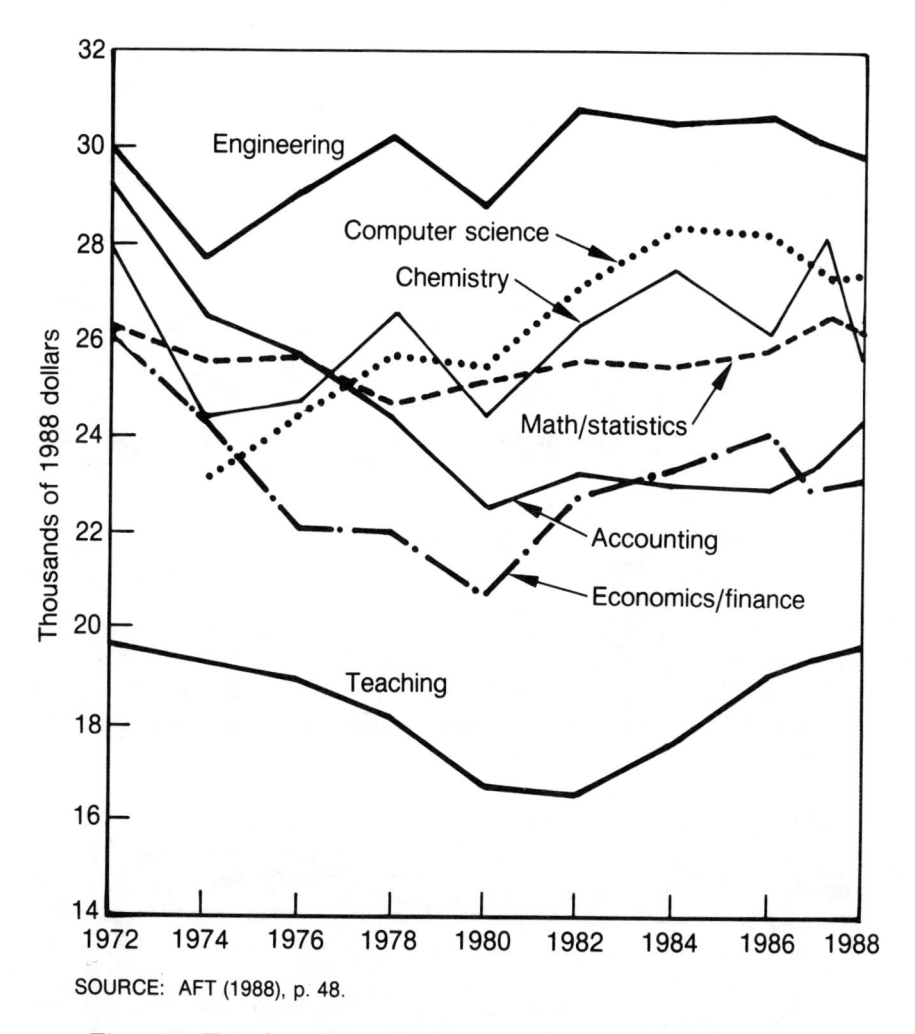

Fig. 4.1—Trends in beginning salaries for college graduates in selected occupations

Table 4.2

RATIO OF EXPECTED SALARIES OF COLLEGE GRADUATES
TO BEGINNING TEACHERS' SALARIES

Field	1972	1974	1976	1978	1980	1982	1984	1986	1987	1988
Teaching	1.00	1.00	1.00	1.00	1.00	1.00	1.00	1.00	1.00	1.00
Engineering	1.52	1.43	1.54	1.66	1.72	1.86	1.73	1.61	1.55	1.52
Accounting	1.49	1.37	1.36	1.34	1.35	1.39	1.30	1.20	1.21	1.24
Chemistry	1.41	1.27	1.31	1.46	1.47	1.59	1.56	1.37	1.45	1.31
Math or statistics	1.33	1.33	1.36	1.35	1.51	1.54	1.45	1.36	1.37	1.33
Economics/finance	1.33	1.26	1.17	1.20	1.24	1.37	1.32	1.26	1.18	1.18
Computer science		1.20		1.41	1.52	1.63	1.61	1.48	1.41	1.39
Others	1.33	1.28	1.30	1.38	1.50	1.51	1.49	1.51	1.18	1.34

SOURCE: AFT (1988), p. 47.

scientists and engineers[1] at the time of each decennial census. The sample is drawn from the census, with follow-up surveys conducted in two to three year cycles over the course of the decade. The most recent sample of individuals was drawn from the 1980 census.

Our analyses of the NSF Surveys of Scientists and Engineers reveal that, since 1970 at least, experienced scientists and engineers have rarely entered teaching.[2] For example, of the 21,423 respondents employed in scientific and technical occupations in 1970, no more than 121 (about 0.5 percent) ever switched to precollege teaching during the course of the decade. Moreover, most of these did not stay in teaching for more than one or two years. Only three of these 121 appear to have worked as teachers over the entire decade.

The NSF data from the 1980s reveal a similar pattern. Among those employed as scientists and engineers in 1980, only about 0.2 percent entered precollege teaching in 1982 or 1984. Further analyses of these data reveal some interesting features of this recruitment pool, and of those in the pool who do enter teaching.[3] As Table 4.3 shows,

---

[1]The target occupations include operations and computer specialists, engineers, mathematical specialists, life scientists, physical scientists, environmental scientists, psychologists, social scientists, and college teachers. Elementary and secondary teachers are not included except as they fall into the "residual" segment of the sample or as individuals move into this occupation over the course of the decade.

[2]A more detailed discussion of the NSF sample, and of our analyses of this sample, is provided in the appendix.

[3]It is important to note that because few scientists entered teaching (52), the results of analyses on this group are not very stable. We therefore discuss only the largest observed differences between those who entered teaching and the total sample; however, we caution the reader that even these differences should be viewed as tentative, subject to verification with larger samples.

the teaching subgroup and the total recruitment pool have some interesting differences and similarities.

First, the two groups differ markedly in their demographic profiles. Scientists as a group tend to be predominantly white and male. In contrast, scientists who enter teaching are much more likely to be female. The teaching group also includes a greater proportion of blacks, though other minorities (principally Asians) are less well represented. This mirrors the greater representation of both blacks and women in teaching generally. Historically, whether for social service or economic reasons, teaching has disproportionately drawn on these groups. The underrepresentation of Asian-Americans is also typical of the overall teaching force (Darling-Hammond, Pittman, and Ottinger, forthcoming).

Teaching entrants in this sample seem to come disproportionately from the 35 to 39 year old age group, suggesting that occupational switches at this career stage are more likely than at other times. Teaching entrants tend to be a bit more highly educated than the sample as a whole. This is partly because of the study required to change fields: 22 percent had obtained (or were in the process of obtaining) degrees in education. Also, a relatively high proportion of those transferring to teaching came from university positions, which typically require at least a master's degree.

Data on fields of study also show that recruits were more likely to have studied the physical sciences and least likely to have studied engineering. Similarly, their occupations in 1980 reveal that relatively few recruits had been employed in engineering (compared to the preponderance of engineers in the overall sample), whereas a relatively high proportion were natural scientists. Postsecondary teachers, social scientists, and computer programmers were also overrepresented in the teaching group. Mathematicians and computer scientists were underrepresented. This makes sense on a financial basis: Wage differentials are lower between the former fields and teaching than for engineering or mathematical and computer sciences.

Table 4.4 compares the 1982 and 1984 salaries of the total sample of scientists with those who entered teaching. Teaching entrants for both years are divided into those teaching in that year and those not teaching in that year, thus indicating their teaching salary and alternative occupation salary (since most stayed in teaching for only one or two years). The table shows that those who enter teaching earn much less in both teaching and their science occupations than do other scientists. Since the two groups do not differ much on average in age or educational attainment, these salary differences do not reflect age- or education-related differences. They may, however, reflect race and sex

Table 4.3

SELECTED CHARACTERISTICS OF THOSE ENTERING TEACHING
AND THE TOTAL NSF SAMPLE: 1982, 1984
(In percent)

Selected Characteristics	Those Entering Teaching	Total Sample
Sex		
Male	57.7	89.9
Female	42.3	10.1
Race		
Black	20.7	2.2
White	78.0	91.0
Other	1.3	6.8
Age (as of 1982)		
<30	7.2	15.1
30–34	11.7	17.5
35–39	42.1	16.7
40–44	10.9	12.0
45–49	9.7	10.3
50–54	5.6	9.4
55–59	8.8	9.0
60+	4.0	10.0
Years of postsecondary education		
0–3	0.0	10.5
4	34.9	36.1
5–6	39.2	31.2
7+	26.0	22.2
Most recent study field (1982)		
Biological/agricultural sciences	8.3	6.2
Education	22.0	1.7
Engineering	20.2	50.3
Health fields	0.0	0.6
Mathematical sciences	1.0	9.7
Physical sciences	26.0	10.4
Psychology/social sciences	16.6	8.0
Arts/humanities/other	5.8	13.1
Occupation in 1980		
Engineer, architect, or surveyor	31.2	62.5
Mathematical or computer scientist	1.1	8.9
Natural scientist	34.1	12.3
Postsecondary teacher	11.5	3.9
Social scientist	11.3	6.6
Computer programmer	10.8	5.8
(N)[a]	(52)	(55,405)

[a]Unweighted number of cases. The proportions shown are based on weighted frequencies.

Table 4.4

SALARIES EARNED BY THOSE WHO ENTERED TEACHING
AND BY THE TOTAL NSF SAMPLE

	Those Who Entered Teaching		
Annual Salary	Teaching in Given Year	Not Teaching in Given Year	Total Sample
In 1982	(N=20)	(N=32)	
Less than $10,000	13.9	3.2	1.9
$10,000–$19,999	22.8	30.5	5.1
$20,000–$29,999	40.3	58.7	27.2
$30,000–$39,999	23.0	7.6	37.7
$40,000–$49,999	0.0	0.0	18.0
$50,000 or more	0.0	0.0	9.0
Mean salary	$21,216	$21,341	$33,445
In 1984	(N=39)	(N=13)	
Less than $10,000	9.4	9.3	1.3
$10,000–$19,999	25.7	3.4	3.5
$20,000–$29,999	63.6	8.1	16.7
$30,000–$39,999	0.0	48.9	36.1
$40,000–$49,999	1.3	30.3	24.8
$50,000 or more	0.0	0.0	17.6
Mean salary	$19,997	$32,628	$38,292

differences in pay. Whatever the reason, teacher entrants obviously came from the lower-paying end of the scientific occupational spectrum (though those who exited teaching between 1982 and 1984 left for higher paying jobs). Thus, in spite of any benefits teaching may offer in terms of work hours, working conditions, or personal satisfaction, science workers seem to be an unlikely source of sustained personpower for teaching.

Defining all individuals in this sample of science workers as a reserve pool for mathematics and science teaching may not be very realistic. Instead, one can argue that certain subgroups—such as those working in the service sector, those with education degrees, or those involved in teaching or training in their job—should have a higher propensity to enter teaching. Indeed, these subgroups do have a higher proportion of transfers into teaching (between 0.9 and 1.0 percent), though these proportions are still small. Not surprisingly, these subgroups tend to have higher proportions of women and to earn salaries that are somewhat lower than those for the overall sample. Interestingly, about 4 percent of this 1980 sample of scientists had education degrees, but only 0.2 percent were teaching at the K–12 level. Thus, for this group, "defectors" outnumbered entrants to teaching by a ratio of 20 to 1. (A more detailed investigation of these subsamples is included in the appendix.)

This analysis suggests tentative hypotheses about the type of recruits to teaching who might be expected to come from the science workforce. We might expect these recruits to be employed in lower-paying occupations in the service sector or in technical and training positions. We might expect a larger share of potential recruits to be minorities and women and to be in their thirties than is true of the science workforce as a whole. We might also expect there to be relatively few of them, although labor market conditions and incentives could change this. This represents a beginning step in our understanding of the pools from which nontraditional teacher preparation programs might draw.

Unfortunately, because so few scientists have transferred into teaching in the recent past, it is difficult to extensively analyze these data, or to even draw firm conclusions on the simple analyses performed here. Although this brief analysis provides some initial insights about potential recruitment pools, much work still needs to be done to increase our knowledge of which individuals, from those targeted by these programs, are most likely to enter and remain in teaching, and of how more recruits can be encouraged to enter the field.

In what follows we begin this type of analysis by examining those who *enroll* in nontraditional teacher preparation programs. Through this type of examination, we can investigate the characteristics of those who are successfully recruited into mathematics and science teaching (or at least into formal preparation for teaching); we can also learn more about the training programs these individuals enter, their teaching experiences, and their propensity to enter and remain in teaching.

PROGRAM RECRUITS

We surveyed current participants and graduates of nine preparation programs described in Section III. Table 4.5 shows the distribution of the nontraditional recruit sample by program and by recruit status (participant or graduate) at the time of our survey. About 37 percent of the total sample of 481 recruits were in retraining programs, and 27 percent each were in recent B.A. programs and in alternative certification programs. Only 8 percent of the total sample were enrolled in midcareer programs. As we discovered in our earlier program survey, midcareer programs tend to be much smaller than the other types of nontraditional programs.

Table 4.5

NUMBER OF RECRUITS, BY PROGRAM TYPE

Program	Participants[a]	Graduates	Total
Midcareer programs	25	14	39
George Mason University	9	11	20
George Washington University	16	3	19
Recent B.A. programs	29	100	129
University of North Carolina	8	49	57
University of Massachusetts	18	43	61
West Virginia University	3	8	11
Alternative certification programs	102	31	133
Houston Independent School District	0	31	31
Winthrop College	102	0	102
Retraining programs	70	110	180
Los Angeles County	70	72	142
Texas Woman's University	0	38	38
Total	124	357	481

[a]The Houston alternative certification program no longer includes preparation for mathematics and science teaching, and the Texas Woman's University program is no longer a separate Hellman program; thus, only *graduates* from these two program sites were surveyed. Participants of the Winthrop College program work as full-time teachers while enrolled in their preparation program. Thus, later analyses that examine recruits' teaching experiences include these participants along with the graduates of other nontraditional programs.

Demographic Profile

Like the teaching force as a whole, these programs tend to attract more female than male recruits (see Table 4.6). However, the midcareer programs in our sample have a much higher proportion of male recruits (66.7 percent) than the other programs (35 to 40 percent). Although this could be an artifact of the small sample size for these programs, it may also be because they recruit mostly older individuals from male-dominated fields (i.e., the military and mathematics/science occupations). The proportion of females in the other three types of programs (60–64 percent), though comparable to the teaching force as a whole, is notably higher than that typically found among secondary school mathematics and science teachers. Nationally, about half of all secondary-level mathematics teachers are female, whereas only about 41 percent of grade 7–9 science teachers and 31 percent of grade 10–12 science teachers are female (Weiss, 1987).

Table 4.6

DEMOGRAPHIC PROFILE OF RECRUITS, BY PROGRAM TYPE
(In percent)

Selected Characteristics	Mid-career	Recent B.A.	Alternative Certification	Retraining	Total
Sex					
Female	33.3	59.7	60.8	64.4	59.7
Male	66.7	40.3	39.2	35.6	40.3
Age					
Younger than 30	12.8	79.8	51.1	8.3	39.7
30–40	46.2	18.6	35.3	48.9	36.8
41–50	20.5	1.6	9.8	35.6	18.1
Over 50	20.5	0.0	3.6	7.2	5.4
(Mean age)	(38)	(27)	(32)	(40)	(33)
Ethnicity					
White	84.6	92.1	80.0	71.1	80.2
Black	10.3	3.9	16.9	7.8	9.6
Hispanic	2.6	.8	1.5	11.1	5.0
Other	2.6	3.1	1.5	10.0	5.2
Occupation before program entry					
Teacher (grades K-12)	5.3	8.1	15.8	95.8	42.8
Postsecondary instructor	2.6	2.4	3.1	0.0	1.7
Science/engineering	57.9	15.3	28.4	0.0	16.6
Other fields	21.1	16.9	18.9	1.0	11.9
Student	2.6	51.6	10.2	1.2	17.3
Other	10.5	5.6	23.6	1.8	9.6
(N)	(39)	(129)	(133)	(180)	(481)

Recruits' age distribution is mostly dictated by the program's recruitment pool. For example, almost 80 percent of recent B.A. program recruits are younger than 30, whereas over 80 percent of mid-career program recruits and retrainees are over 30. Just over half of the alternative certification program recruits are under 30.

Except for the recent B.A. recruits, these nontraditional recruits tend to be older than traditional recruits to teaching. Overall, 60 percent of the nontraditional recruits are over 30; their average age is 33. In comparison, three-quarters of all college students (the "traditional" recruitment pool for teachers) are younger than 30; 60 percent are younger than 25 (U.S. Bureau of the Census, 1987). At the same time, only a small fraction (5 percent) of the nontraditional recruits are prospective retirees (defined as those over the age of 50). As we found in our survey of programs, this pool of prospective recruits seems limited.

These nontraditional programs also appear to be attracting substantially more minority candidates than other teacher preparation programs. Overall, 20 percent of the nontraditional recruits are minority group members; 10 percent of the new recruits are black. By comparison, the 1985 Survey of Recent College Graduates found that only 9 percent of all bachelor's and master's degree candidates newly qualified to teach (in all subject areas) were minority group members; 5.6 percent were black (CES, 1986).

Though minority enrollments in recent B.A. programs are roughly the same as these national norms, the other program types have substantially higher minority enrollments, including the intense and costly midcareer programs. The high proportions of minority enrollment among these particular programs may be partially due to their location in metropolitan areas with ethnically diverse populations (e.g., Washington, D.C., and Los Angeles); we cannot say whether these programs are attracting more minority candidates than other programs in those areas. Whether because of location, financial aid, or active recruitment, these programs appear to be successful at recruiting minority candidates. Given the current shortage of entering minority teachers and the underrepresentation of minorities in mathematics and science teaching,[4] these findings are very encouraging and should be further pursued.

Employment and Educational Backgrounds

We also asked recruits about their main activity the year before entry into the program. Obviously, the overwhelming majority of the retrainees were previously employed as teachers in grades K-12. Eleven percent of the nonretrainees had been employed as elementary or secondary teachers, including about 16 percent of the alternative certification recruits, many of whom had been teaching in private schools. Over half (58 percent) of the midcareer recruits had been working in a scientific or engineering field; just over one-quarter of those in the alternative certification programs were also working in these fields. This contrasts with the recent B.A. program recruits, over half of whom were students in the year before entry. Further analyses of the "other" category revealed that only 0.5 percent of all recruits

---

[4]Currently, only 8 percent of all 7-12 mathematics teachers, and 10 percent of all 7-12 science teachers are minority group members (Weiss, 1987).

were retirees, 3.3 percent were unemployed, and 3.8 percent were homemakers. Clearly, these latter groups have not been large recruitment pools for these nontraditional programs.

Those who were not teaching in elementary/secondary schools in the year before entering the program were asked about their most recent *permanent* occupation. A small proportion of those in the alternative certification programs and almost half of the recent B.A. program recruits had had no permanent job before program entry; among the remaining sample, 85.7 percent listed an occupation. About 50 percent of these recruits had been working in the private sector before joining the program, and about 20 percent were employed by universities or colleges. A smaller proportion, about 17 percent, were drawn from the government sector (including 8 percent from the Armed Forces).

Recruits' previous jobs were fairly evenly split between the professional and technical/support fields (Table 4.7). The midcareer programs appear to be most successful in drawing from the managerial/professional specialties pool, their primary target pool. Forty percent of the recruits to these programs were scientists or engineers, more than double the proportion in any of the other types of programs; a large proportion of the engineers entering midcareer programs came from the military. The much larger alternative certification and recent B.A. programs drew more of their recruits from administrative support and service occupations. Overall, these lower-paying fields provided the larger number of nontraditional program entrants.

Tables 4.8 and 4.9 present data on the educational attainment of participants and graduates, respectively, and their major fields of study for their undergraduate and graduate degrees.[5] All participants have a bachelor's degree. Retrainees, not surprisingly, had majored primarily in education and nonscience fields; most other recruits had majored in mathematics or science. Over half of the retrainees have at least a master's degree; this is at least partly attributable to the fact that retrainees tend to be somewhat older, and to the incentives given teachers to obtain a master's degree.[6] About a third of the nonretrainee program recruits (primarily those in the midcareer programs) have higher degrees.

---

[5]Participant and graduate data are reported separately because graduates of the following four programs receive a master's degree in teaching or education upon program completion: George Washington University, University of North Carolina, University of Massachusetts, and West Virginia University. This includes all of the recent B.A. programs and one of the two midcareer programs.

[6]Most teaching salary schedules include increments for higher degrees and coursework. Some states also require that teachers earn a master's degree to maintain their teaching license.

Table 4.7

RECRUITS' MOST RECENT PERMANENT OCCUPATION BEFORE PROGRAM ENTRY, BY PROGRAM TYPE[a]

Occupation[b]	Mid-career	Recent B.A.	Alternative Certification	Total
Managerial and professional specialties	67.6	30.2	41.5	43.5
Managers/administrators	5.4	5.7	10.6	8.2
Engineers	16.2	1.9	6.4	7.1
Mathematicians/scientists	24.3	5.7	9.6	11.4
Social workers/counselors	0.0	7.5	3.2	3.8
Health field workers	5.4	1.9	4.3	3.8
Postsecondary teachers	5.4	3.8	4.3	4.3
Others	10.8	3.8	3.2	4.9
Administrative support	16.2	58.5	43.6	42.4
Technicians/research assistants	13.5	32.1	12.8	18.5
Health-related technicians	0.0	5.7	12.8	8.2
Others	2.7	20.7	18.1	15.8
Service occupations	0.0	7.5	6.4	5.4
Production, craft, and repair occupations	0.0	1.9	3.2	2.2
Armed Forces	21.6	1.9	6.4	8.2
(N)	(37)	(53)	(94)	(184)

[a] Ninety-six percent of those in the retraining programs were elementary or secondary teachers. These programs have therefore been omitted from this table.

[b] Categories based on 1980 census occupation codes. Columns may sum to greater than 100, as those in the Armed Forces were also classified by occupation whenever possible.

Program graduates' education is similar, except that more have master's degrees (see Table 4.9). All recent B.A. programs in this study grant a master's degree to their graduates; thus, over 90 percent of those graduating from the recent B.A. programs have a master's degree in education. (A few graduates reported they had not yet received their degrees, possibly because of missing requirements or other administrative problems.)

Table 4.8

EDUCATIONAL BACKGROUND OF PARTICIPANTS, BY PROGRAM TYPE

Educational Background	Nonretraining Programs[a]	Retraining Programs
Bachelor's degree	100.0	100.0
Education	0.0	36.8
Mathematics/science	81.5	5.9
Other fields	22.3	63.2
Master's degree	25.9	53.6
Education	0.0	34.8
Mathematics/science	13.0	1.4
Other fields	14.9	18.8
Ph.D.	5.6	1.4
Professional degree	1.9	0.0
(N)	(54)	(70)

NOTE: Major fields columns may sum to greater than 100 percent because some participants had second degrees.

[a] Includes midcareer, recent B.A., and alternative certification programs. There were no significant differences among participants in these groups; hence the data are aggregated.

Table 4.9

EDUCATIONAL BACKGROUND OF GRADUATES, BY PROGRAM TYPE

Educational Background	Mid-career	Recent B.A.	Alternative Certification	Retraining
Bachelor's degree	100.0	100.0	100.0	100.0
Education	0.0	4.0	0.0	33.6
Mathematics/science	92.8	87.0	86.5	12.7
Other fields	7.1	10.0	13.5	53.7
Master's degree	42.9	94.0	27.8	41.8
Education	7.1	92.0	6.0	26.4
Mathematics/science	21.3	13.0	14.3	3.6
Other fields	21.4	4.0	8.3	14.5
Ph.D.	21.4	0.0	4.5	1.8
Professional degree	0.0	1.0	0.8	1.8
(N)	(14)	(100)	(133)	(110)

NOTE: Major fields columns may sum to greater than 100 percent because some participants had second degrees.

Prior Teaching Experience

One characteristic of recruits that is of considerable interest is their prior teaching experience. How many have taught before? In what capacity? And where? Answers to these questions are important in sketching yet another dimension of the reserve pool, as well as in determining the level of information these recruits are likely to have about the "experience" characteristics of teaching (defined above as characteristics of the teaching job that can be learned only through actual experience).

Forty-three percent of this sample of new recruits (mostly those in retraining programs) were teachers in the year before program entry. An additional measure of prior teaching experience is certification status before program entry, as becoming certified through traditional routes usually involves at least some student teaching. Table 4.10 shows the certification status of recruits. Perhaps it is not surprising, given the recent shortages of mathematics and science teachers, that about 14 percent of those who reported teaching in the year before program entry had never been certified to teach. In many cases, these were retrainees or alternative certification candidates currently teaching mathematics or science with emergency credentials, or teachers from private schools. As expected, over 90 percent of nonteachers had never been certified, although 11 percent had taught previously.

To find out about other teaching experiences these recruits may have had, recruits who were not classified as teachers in the prior year were asked about any teaching position they had held, including the subjects they taught, and by whom they were employed. Those who *were* teaching in the previous year were asked about their main teaching assignment. These data are summarized in Tables 4.11 and 4.12.

Table 4.10

TEACHING CERTIFICATION STATUS OF RECRUITS, BY OCCUPATION BEFORE PROGRAM ENTRY

Certification Status	Teacher	Student	Science Field	Other
Never certified	14.1	91.0	95.9	91.5
Certified in the past, but not at entry	3.1	0.0	2.7	3.8
Certified at entry	82.8	9.0	1.4	4.7
(N)	(191)	(78)	(74)	(106)

Of the nonteachers, three-quarters had had some kind of teaching experience (Table 4.11). Former students are the most experienced group, with 85 percent of these recruits having some type of teaching experience, generally as a tutor or teaching assistant at college. Two-thirds of entering science-field workers had taught as tutors or instructors, often in private industry. Workers from nonscience fields were more likely to have worked as postsecondary instructors. As noted above, about 10 percent of the recruits coming from jobs other than elementary or secondary teaching have worked in these occupations in

Table 4.11

RECRUITS' PRIOR TEACHING EXPERIENCE, FOR THOSE WHO DID NOT TEACH IN PRIOR YEAR, BY OCCUPATION BEFORE PROGRAM ENTRY

Teaching Experience	Student	Science Field	Other Fields	Total
Percentage with any teaching experience	85.0	67.5	75.0	75.8
Position held				
Tutor	69.7	45.7	38.2	51.1
Instructor	4.5	23.9	22.4	16.5
Postsecondary instructor	7.6	17.4	32.9	20.2
Elementary/secondary teacher	4.5	10.9	13.2	9.6
Teacher aide	10.6	2.2	15.8	10.6
Teaching assistant	27.3	23.9	27.6	26.6
Other	27.3	28.3	32.9	29.8
Employer				
Elementary school	6.3	0.0	10.0	6.1
Secondary school	17.5	17.3	21.0	18.9
College/university	71.4	28.8	46.9	50.0
Private industry	27.0	53.8	37.0	38.3
Other	19.0	42.3	32.1	30.6
Subject taught				
Biological/life sciences	24.1	35.1	25.7	27.3
Chemistry	18.5	18.9	10.0	14.9
Computer science	9.3	16.2	5.7	9.3
Earth/space science	1.9	10.8	5.7	5.6
General science	3.7	16.2	14.3	11.2
Mathematics	63.0	27.0	32.9	41.6
Physics	7.4	16.2	14.3	12.4
Other	24.1	40.5	52.9	40.4
(N)	(68)	(52)	(81)	(201)

the past. Close to 20 percent had been employed in secondary schools in some capacity—as teacher, aide, or tutor.

Colleges and universities are the most common employers of those with teaching experience (50 percent), with private industry being the next most common employer (38 percent). Mathematics is the subject most commonly taught by these recruits, whereas earth science and computer science are the subjects taught least frequently (6 and 9 percent, respectively). Only 15 percent of these recruits have experience teaching chemistry, and only 12 percent have taught physics.

Teachers in retrainee programs come from a wide variety of teaching assignments (Table 4.12), with the largest proportions coming from elementary teaching (27.5 percent) and mathematics or computer science (23.5 percent). Nearly one-third had been teaching science or mathematics as their main assignment before entering their retraining program. Presumably, their retraining is intended to improve their capacities or credentials to teach in these areas. Eighty-seven percent of the former teachers enrolled in other types of programs had had mathematics or science as their main assignment area while teaching.

Table 4.12

PRIOR TEACHING ASSIGNMENT OF FORMER TEACHERS, BY RECRUIT TYPE

Main Assignment	Retrainee	Nonretrainee	Total
Mathematics/computer science	23.5	25.9	24.2
Science	7.8	61.1	21.7
Elementary education	27.5	3.7	21.3
Business/vocational education	7.8	1.9	6.3
Special education	7.2	1.9	5.8
Physical education/health	5.9	0.0	4.3
Social science	5.2	0.0	3.9
English/language arts	4.6	1.9	3.9
Fine arts	4.6	0.0	3.4
Other	5.9	3.7	5.3
(N)	(153)	(54)	(207)

REASONS FOR ENTERING MATHEMATICS AND SCIENCE TEACHING

We turn now to an examination of the factors that motivate these individuals to enter mathematics and science teaching. There are, as one would expect, both differences and similarities in the reasons ranked highly by different types of recruits (see Table 4.13). The most striking finding is how differently teachers as compared to other recruits rank almost all of these factors. For the most part, this difference can be attributed to the fact that teachers are answering a different question than the other recruit groups. Having already chosen teaching as an occupation, teachers are explaining why they chose to switch to *mathematics or science* teaching. Thus, for these recruits, current high demand in these areas and job security are more important factors than they are for other recruit types, whereas more idealistic reasons for choosing teaching (working with young people, etc.) are relatively less important. These factors may have motivated them earlier to choose teaching as a career, but they are less important reasons for switching to a new specialty area. The importance of labor market demand and job security suggests that many of these teachers may feel vulnerable to layoffs if they do not become qualified to teach a high-demand field.

Across all recruit types, interest in the subject-matter field is rated as the most important reason for entering mathematics or science teaching. For nonteachers, this is followed by the opportunity to work with young people, the belief that their abilities are well-suited to teaching, and a desire to contribute to the betterment of society. Not surprisingly, financial considerations are ranked as the least important reasons for entering teaching; overall, only 6 percent of all recruits were attracted to mathematics or science teaching because of its financial rewards.

In comparison, teachers surveyed by the National Education Association (1987) ranked (a) the desire to work with young people, (b) value or significance of education in society, and (c) interest in subject-matter field as the three most important reasons in their original decision to become a teacher. "Long summer vacations" were also cited fairly frequently, but salary or other financial rewards were not often selected. Thus, these nontraditional recruits appear to mirror the motivations of most teachers to a large extent, although for these recruits, interest in the subject area appears to be a stronger motivation for entering mathematics or science teaching than it is for teachers in general.

Table 4.13

REASONS FOR INTEREST IN MATHEMATICS AND SCIENCE TEACHING, BY OCCUPATION BEFORE PROGRAM ENTRY

Reasons	Teacher	Student	Science Field	Other	Total
Interest in subject-matter field	91.1	91.3	84.4	86.1	88.8
Abilities are well-suited to teaching	48.2	68.8	64.9	69.4	59.6
Opportunity to work with young people	35.1	73.8	71.4	73.1	56.6
Current high demand in this field	60.7	41.3	57.1	46.3	53.3
Contributes to betterment of society	30.4	60.0	58.4	52.8	45.6
Good vacation time	25.1	56.3	57.1	59.3	44.1
Provides opportunity to be creative	28.8	43.8	46.8	38.0	36.6
Job security	41.1	16.3	23.4	29.6	31.1
Good working hours	20.4	32.5	33.8	40.7	29.6
Encouragement from others I respect	13.6	22.5	26.0	27.8	20.6
Provides a change from my past work	7.9	2.5	48.1	30.6	19.1
Provides added income	6.3	5.0	13.0	10.2	8.1
Good salary	3.1	0.0	9.1	14.8	6.4
(N)	(191)	(80)	(77)	(108)	(456)

COSTS OF PARTICIPATION

One important characteristic that determines who will enroll in these nontraditional preparation programs is the economic cost of program participation. The costs of training or retraining for an occupation include (a) explicit costs, in terms of tuition and other expenses, (b) forgone earnings while training, and (c) the value of the time individuals sacrifice to attend the program.

As seen previously, program duration is quite variable, ranging from 16 weeks to three years, and time costs are equally wide-ranging. Full-time programs usually take 12 to 15 months to complete.[7] Most part-time programs take between two to three years to complete. Part-time programs, however, allow recruits to maintain their earnings. Alternative certification programs allow participants to start paid teaching jobs almost immediately; some programs that provide internships also provide participants with earnings. Thus, in many cases, time costs are minimized or offset by earnings opportunities.

There is considerable variation in tuition expenses. Tuition costs range from zero in the alternative certification programs, the Los Angeles County retraining program, and the Lyndhurst recent B.A.

---

[7]An exception is the Switcher Program at GMU, which takes only 16 weeks.

program, to a total tuition of $10,800 for the George Washington University midcareer program. All programs with explicit tuition costs also have some form of student aid to reduce these costs for at least some participants.

As Table 4.14 shows, recruits' methods of financing tuition and other program costs vary across programs. Since recruits were asked to list all sources of funding, these data do not reveal each respondents' predominant mode of funding; however, they do show how funding sources differ among recruits in the different programs. For example, direct government funding (from school districts, counties, or states) is more commonly available to retrainees and those in alternative certification programs, whereas traditional loans and grants are more common among those in midcareer and recent B.A. programs. Recent B.A. program recruits use the widest range of funding sources, with the major sources consisting of fellowships, paid internships, and guaranteed student loans.

Recruits' estimates of their out-of-pocket tuition costs reveal that almost half of all recruits pay no out-of-pocket tuition costs (see Table 4.15). Retrainees are particularly likely to have no tuition costs (84 percent), or to have only low costs; alternative certification program recruits also tend to have either no costs or moderately low costs. Just over half of recent B.A.s have no out-of-pocket tuition costs, although one-third pay over $2,000. Midcareer programs appear to be most

Table 4.14

RECRUITS' METHODS OF FINANCING PROGRAM COSTS,
BY PROGRAM TYPE

Method of Financing	Mid-career	Recent B.A.	Retraining	Alternative Certification
Guaranteed student loan	2.6	27.1	0.6	4.5
Other loan	20.5	7.0	0.6	0.8
Fellowship	0.0	45.0	0.6	0.0
Scholarship	5.1	1.6	2.2	0.0
Grant	30.8	0.8	17.8	0.8
Paid internship	0.0	38.0	1.1	12.0
Your own/family income	74.4	45.7	15.0	54.1
School district funding	2.6	1.6	11.1	24.1
County funding	0.0	0.0	57.8	0.8
State funding	15.4	1.6	19.4	68.4
(N)	(39)	(129)	(180)	(133)

costly to participants: Although over one-quarter of participants pay nothing, nearly the same proportion pay over $4,000. These more costly programs are also the smallest programs.

An additional measure of opportunity costs is forgone earnings, or salary earned in the occupation before entering the teacher preparation program. For some individuals, these costs are likely to be zero (those not working, for example, or those attending the program part-time while continuing to work). In fact, excluding former teachers, only 58 percent of these nontraditional recruits ever held a permanent job before program entry. For others, however, these costs could be substantial.

The large economic costs facing individuals considering leaving the scientific field are highlighted in Fig. 4.2, which compares teacher salaries in 1985–86 to the 1984 salaries earned by those scientists and engineers represented in the National Science Foundation Survey of Scientists and Engineers. These two sets of data, based on samples of comparable ages and educational levels, demonstrate that teachers earn, on average, significantly less than those working in the scientific and engineering fields.[8] In 1985–86, teachers earned $24,504 on average. Additional earnings from summer employment and other stipends raise this to $27,780. This contrasts sharply with the 1984 average salary of scientists and engineers, which was $38,292. Although only 5 percent of scientists earned less than $20,000, 25 percent of teachers made less than this amount. On the other hand, 78.5 percent of

Table 4.15

TOTAL OUT-OF-POCKET TUITION COSTS FOR PROGRAM RECRUITS, BY PROGRAM TYPE

Out-of-Pocket Tuition Costs	Mid-career	Recent B.A.	Retraining	Alternative Certification
$0	28.6	51.9	83.9	33.9
$1–500	8.6	2.9	12.9	13.0
$501–1000	17.1	5.8	1.3	51.3
$1001–2000	8.6	3.8	1.9	1.7
$2001–4000	14.3	22.1	0.0	0.0
$4000+	22.9	13.5	0.0	0.0
(N)	(35)	(104)	(155)	(115)

---

[8]This comparison actually underestimates the wage gap between the two occupational fields, as scientists in the 1984 NSF survey would have had increased earnings in the following year.

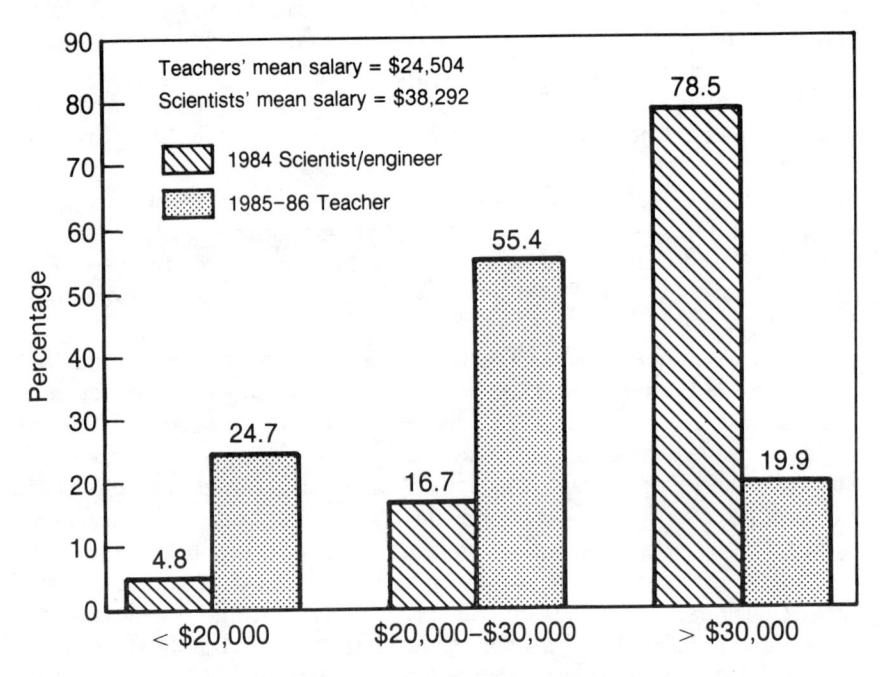

Fig. 4.2—Comparison of salaries of scientist/engineers and teachers

scientists made at least $30,000, whereas only 20 percent of teachers made this amount.

These large differences in salary are not, however, reflected in the data we collected on program recruits. As Fig. 4.3 shows, the salaries of recruits *entering teaching* from the science fields do not appear so extremely different from those of teachers. In this sample, 37 percent of those who were teachers when they entered the programs and 44 percent of transferring scientific workers earned less than $20,000. Seventeen percent of teachers and 30 percent of transferring scientists earned $30,000 or more in their previous job. Recruits from scientific fields who entered the teacher preparation programs in our sample earned disproportionately lower salaries than scientists overall. This finding is consistent with our analysis of those who entered teaching from the NSF sample; these individuals also earned significantly less than scientists in general. This is also consistent with our finding that half of these recruits come from technical and support jobs rather than the higher-paying scientific occupations, and further reinforces the hypothesis that individuals with lower opportunity costs are more likely to enter teaching from other occupations.

This hypothesis is also supported by the salary data on those from nonscience, nonteaching fields ("other" fields), also shown in Fig. 4.3. Seventy-one percent of these individuals earned less than $20,000 in their most recent full-time occupation, whereas only 14 percent earned at least $30,000. For these individuals, therefore, a change to teaching may often lead to a salary *increase*.

SUMMARY

Recruits in nontraditional teacher preparation programs contain a higher than average representation of minority, female, and older candidates for mathematics and science teaching. They also bring to teaching a wide range of backgrounds and experiences, including experiences in scientific and nonscientific fields, bachelors' degree subject-area training, and a range of past instructional experiences. Those recruits who do come from a scientific working background are more likely to come from the lower-paying technical, support, and service fields than they are from the professional and managerial fields. Those coming from nonscience fields are also drawn disproportionately from jobs in the lower salary ranges.

Although teachers in retraining programs are more likely to have a master's degree than are other new recruits, the fact that many of the recent B.A. and midcareer programs grant master's degrees means that after program completion, most of these new recruits enter teaching at the master's degree level. Just over 20 percent of midcareer program recruits also have a doctoral degree. Thus, many of these new recruits are entering teaching with a significant amount of both education and subject-area training.

Most of these new recruits (in addition to the retrainees) also have prior teaching experience, typically as teaching assistants, instructors, or tutors. This prior experience is probably a good indicator of recruits' interest in teaching in general, and suggests that these individuals have some degree of experience with basic pedagogical issues, such as organizing and pacing course material, explaining concepts, and evaluating students' work. However, as will become apparent in later analyses, program participants are by no means fully prepared—either in terms of pedagogical knowledge or more general expectations—for entering the K–12 classroom.

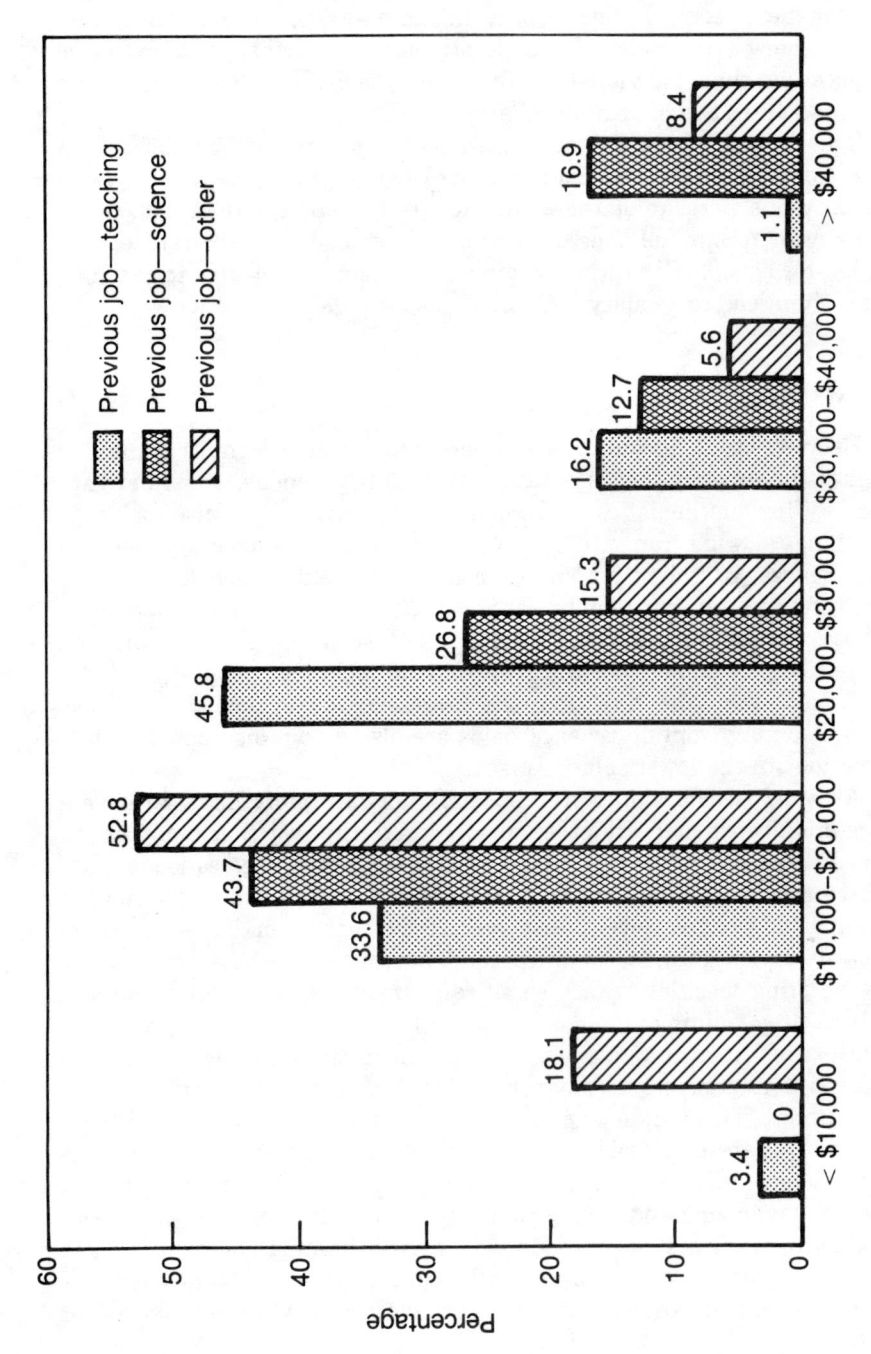

Fig. 4.3—Comparison of salaries earned by nontraditional recruits in previous jobs

The recruits are fairly consistent in their reasons for their interest in mathematics and science teaching, and greatly resemble "traditional" teachers in their interest in subject matter, children, and contributing to society. Both traditional and nontraditional recruits appear to perceive teaching as an occupation that offers few financial rewards, but many personal and social rewards.

Finally, recruits use a variety of financial methods to cover program costs. Many take advantage of fellowships, grants, and school district, county, or state funding, and many have paid teaching positions or internships while enrolled in their program. Except for those in retraining programs, however, most must supplement these funds with their own income. The amount of out-of-pocket tuition costs the recruits pay also varies widely by program type, with midcareer programs being the most costly to participants. For midcareer and recent B.A. program recruits, there appears to be a bimodal distribution for participants' costs—some participants have no, or very low, costs, and others face substantial costs for their teacher training.

V. NEW RECRUITS' PROGRAM EXPERIENCES

The perceptions and opinions of those who enroll in a teacher preparation program provide a rich source of data for examining how the program achieves its goals. This section summarizes the program experiences of new recruits, including their reasons for selecting particular teacher training programs, the nature of their teaching practicum, their evaluation of program components, perceptions of program strengths and weaknesses, and recommendations for program improvement. These data do not constitute a program evaluation, but they shed light on the unique needs of nontraditional recruits and on recruits' perceptions of how well existing programs address these needs.

PROGRAM ATTRACTIONS

Although new recruits choose to enter mathematics or science teaching for mainly altruistic and intellectual reasons, practical reasons dominate their choice of particular teacher preparation programs (Table 5.1). Financial assistance, through either tuition assistance or paid internships, is an important incentive for attracting recruits; pragmatic considerations that make program attendance more feasible are also valued program features. These latter features include the compatibility of the program with other time commitments, its duration and location, and the award of credit for past experience and education. These considerations seem to override others, such as the nature of coursework and other training features or characteristics of the institution. These substantive concerns tend to become more important in determining recruits' later satisfaction, rather than in recruitment.

Typically, recruits could not choose among a wide range of nontraditional programs—rather, their choice was likely to be limited to one, or perhaps two, programs that were readily available and were structured to make teacher preparation a feasible goal. Nonetheless, these program attributes may have occasionally motivated recruits' choices among several program options, and in many cases these features may have persuaded recruits that preparing to teach was worth trying.

Table 5.1

PROGRAM FEATURES INFLUENCING RECRUITS' APPLICATION
DECISIONS, BY PROGRAM TYPE
(In percent)

Program Features	Mid-career	Recent B.A.	Alternative Certification	Retraining
Short, intensive program	61.5	86.0	67.7	50.0
Paid internship	NA	48.8	41.4	NA
Compatibility of program with other time commitments	51.3	11.6	31.6	76.7
Tuition assistance	38.5	47.3	14.3	61.7
Location of institution/program	69.2	56.6	9.8	51.1
Acceptance of prior education and job experience	56.4	13.2	65.4	32.2
Academic courses offered	2.6	22.5	2.3	35.0
Prestige of institution/program	28.2	33.3	4.5	11.1
Education courses offered	20.5	8.5	9.8	13.3
Academic caliber of students	10.3	23.3	2.3	3.9
Placement assistance	10.3	14.0	4.5	5.0
Supervised field training	7.7	10.9	11.3	2.8
(N)	(39)	(129)	(133)	(180)

RECRUITS' PERCEPTIONS OF PROGRAMS

Recruits were asked a variety of questions about the perceived quality and effectiveness of their preparation program, both in terms of program components (coursework and practicum), and the supervision or assistance received during their practicum. These ratings were supplemented by open-ended questions asking for recruits' opinions of program strengths and weaknesses and their recommendations for improving the program.

Ratings of Coursework

Table 5.2 presents recruits' evaluation of their program coursework and practicum experience. These tabulations are difficult to interpret definitively as they may reflect the perceived need for or utility of particular types of knowledge or they may reflect the perceived quality of coursework offered in particular programs. Open-ended responses reported in the next section help to explain the ratings assigned by recruits. Some general findings, though, are apparent.

Table 5.2

RECRUITS' EVALUATION OF PROGRAM COMPONENTS, BY PROGRAM TYPE
(In percent)

Program Components	Midcareer Not Valuable	Midcareer Highly Valuable	Recent B.A. Not Valuable	Recent B.A. Highly Valuable	Alternative Certification Not Valuable	Alternative Certification Highly Valuable	Retraining Not Valuable	Retraining Highly Valuable
Education coursework								
Pedagogy/teaching methods	0.0	47.4	2.8	36.9	6.1	6.1	4.9	38.9
(N)	(38)		(141)		(99)		(141)	
Educational theory	10.5	34.2	12.1	16.1	9.1	3.0	9.6	30.4
(N)	(38)		(124)		(99)		(115)	
Child development/learning	0.0	50.0	4.5	26.1	3.0	11.0	16.2	26.5
(N)	(38)		(134)		(100)		(68)	
Subject area coursework	0.0	41.2	18.1	39.3	4.2	48.6	0.6	73.1
(N)	(17)		(61)		(72)		(171)	
Teaching practicum	0.0	73.9	1.6	53.6	3.1	43.3	N.A.	N.A.
(N)	(23)		(125)		(97)		—	

NOTE: Ratings based on a 1–5 scale, with "1" labeled "not at all valuable," and "5" labeled "extremely valuable." The "highly valuable" rating shown here includes both "4" and "5" responses.

First, as a group, the alternative certification recruits appear to be the least satisfied with their coursework; midcareer transfers appear to be the most satisfied. Since alternative certification programs tend to deemphasize the importance of coursework in favor of supervised field experience, they offer less of it. Given the minimal amount of time devoted to education coursework in these programs, it is perhaps not surprising that recruits do not find their courses highly valuable. Except for those in the alternative certification programs, other recruits were much more likely to find their education coursework valuable than not.

Interestingly, in our sample the midcareer transfers—who were the most pleased with their coursework—were also the only recruit type (other than retrainees) who had all received complete pedagogical training before entering the classroom as practicing teachers. (None of the alternative certification program recruits nor the MESTEP recruits have completed their coursework when they begin teaching.) The prior completion of pedagogical coursework may make it easier for recruits to absorb and then integrate teaching knowledge into their teaching practice; it is clearly more difficult to "learn while doing" than it is to apply prior knowledge to one's current work.

Within each group, courses in pedagogy and teaching methods tended to outrank courses in education theory for their perceived value. However, current teachers (retrainees) were more apt to find theory courses valuable. This is consistent with one view held by some teachers and teacher educators that an iterative approach to learning theory and its applications is desirable, that is, theory may inform practice but practice also enriches the understanding of theory as one gains experience against which to apply it.

It is also interesting to note that nonretrainees view subject area coursework to be about as valuable as education coursework, even though these individuals are presumably already knowledgeable in their subject area. This suggests that the common assumption that programs can focus on either one of these areas to the exclusion of the other (depending on recruits' prior background) may be misleading.

By and large, recruits found their coursework to be moderately valuable. Except for the three-quarters of retrainees who found their subject area coursework highly valuable, no more than half of any group gave a four-star endorsement to any group of courses they took. Although these data may not appear overwhelmingly positive, they are comparable to other evaluations of education coursework. For example, Loadman et al. (1988), in a study of Ohio State University education students and graduates, found that only about one-third of all teachers and student teachers considered their education coursework to be of good quality. Pigge (1978) also found that teachers did not feel their education training was as helpful as on-the-job training (i.e., supervised experience) in developing the most highly needed teaching proficiencies.

The Teaching Practicum

All nontraditional programs, except those for retrainees, place great emphasis on the teaching practicum as a crucial learning experience for new recruits. For many of these recruits, the practicum provides their first exposure to the school environment in many years; some programs even encourage participants to observe in schools before program entry, or before beginning their practicum, to more fully prepare recruits for their initial entry into the school environment.

Information on recruits' practicum experience is provided in Table 5.3. The vast majority of recruits serve their practicum in grades 7–12; over 80 percent work with students at the high school level (grades 9–12). (Some work with students at more than one level during the course of their practicum.) These practicum placements appear to match recruits' eventual or planned teaching positions. Most recruits (85 percent overall) are given complete classroom responsibility at some point during their practicum. For alternative certification candidates this is the main form of their practicum; for others a variety of roles are assumed, implying graduated classroom responsibility.

Many recruits have their practicum supervised by more than one individual. For most of those in midcareer or recent B.A. programs, assistance is provided by both cooperating teachers and university supervisors. School principals and other district or program personnel are much less likely to supervise or assist recruits in these programs, doing so in less than 20 percent of all cases. The pattern for alternative certification recruits is different, because these recruits are actually employed as regular teachers in their first year. They are less likely to be assigned a cooperating teacher. Thus, among this group of recruits, school principals and other district personnel are most likely to be charged with providing assistance or supervision.

The amount of time recruits meet with their supervisors each week also varies substantially among program types, with most midcareer program recruits spending two or more hours each week with their supervisors, whereas most recent B.A. and alternative certification recruits spend less than one hour a week with their supervisors. The greater time midcareer recruits spend with supervisors may be a result of the smaller size of these programs, which may allow for more careful supervisor selection, a lower supervisor load, or more supervisors per

Table 5.3

CHARACTERISTICS OF RECRUITS' TEACHING PRACTICUM, BY PROGRAM TYPE[a]

Practicum Characteristics	Mid-career	Alternative Certification	Recent B.A.
Percentage of Recruits			
Grades taught			
Elementary (K–6)	4.0	6.2	.9
Middle (7–8)	8.0	24.8	30.8
Secondary (9–12)	92.0	80.5	86.3
Levels of responsibility			
Complete	80.0	92.0	79.8
Shared	52.0	10.6	45.4
Served as aide	12.0	1.8	18.5
Other arrangement	8.0	2.7	13.4
Practicum supervisor			
Cooperating teacher	100.0	48.2	81.0
Principal	16.0	78.1	19.0
Other district personnel	16.0	48.2	9.5
University supervisor	64.0	10.5	75.9
Other university faculty	12.0	3.5	9.5
Other program students	4.0	4.4	6.0
Hours per week spent with supervisors			
0–1 hours	24.0	67.5	64.0
2–3 hours	32.0	16.7	21.9
3–4 hours	12.0	1.8	8.8
4–5 hours	16.0	2.6	2.6
5 or more hours	16.0	11.4	2.6
Average Hours per Week			
Teaching only	18.9	26.9	15.6
All instructional activities	45.6	44.6	33.6
(N)	(25)	(115)	(119)

[a]Retrainees are not included in this analysis, as these practicing teachers do not have a practicum as part of their preparation program.

recruit. Considering the relative emphasis alternative certification programs place on the teaching practicum (as opposed to education coursework), the finding that most are supervised less than one hour a week is very discouraging. The contrast between the supervision afforded by alternative certification programs and others is even more striking given that recent B.A. and midcareer programs carry a minimal teaching load during their internship (teaching only 16 and 19 hours a week, respectively, in contrast with alternative certification candidates' 27 hour a week teaching load), while receiving more assistance.

It is not surprising, then, that the least favorable ratings for the practicum were among the alternative certification recruits. In effect, they were not actually in a supervised practicum, but were employed as teachers, largely unsupervised, while completing their education coursework. These individuals were clearly less likely to receive the amount of attention and support that other recruits received as part of a teaching practicum.

To a large extent, the value of the practicum depends on the type of supervision and assistance recruits receive. As Table 5.4 shows, the most valuable supervision and assistance was that provided by cooperating teachers and by other program students. Virtually all midcareer and recent B.A. recruits had cooperating teachers, but fewer than half of alternative certification recruits did. Advice from university supervisors was also rated as highly valuable by at least half of the midcareer and alternative certification program recruits; recent B.A. program recruits did not appear to find their university supervisors as helpful. (This may be due in part to MESTEP's practicum arrangement, which has recruits serving their practicum in schools that are located one to three hours away from the university.)

University supervisors may also not be as helpful as cooperating teachers for two reasons. First, they are not as familiar with the actual school situation and typical teaching problems as are school teachers. Second, they usually must supervise many student teachers, so that they cannot give each one as much individual attention as can a cooperating teacher. Practicum advice from those even more removed from the teacher preparation program and the classroom environment—school principals, other district personnel, and other university faculty—was not rated highly; about one-quarter of all recruits rated such advice as "not at all valuable." It is fortunate then that midcareer and recent B.A. program recruits most often receive assistance from cooperating teachers and university faculty, rather than from less-valued advisors. Alternative certification recruits, however, are often supervised by school principals and other district personnel, whose advice tends to be rated as less valuable; this may help explain the lower rating given to the value of the practicum by these recruits.

Table 5.4

RECRUITS' EVALUATION OF SUPERVISION AND ASSISTANCE RECEIVED
DURING PRACTICUM, BY PROGRAM TYPE
(In percent)

	Midcareer		Recent B.A.		Alternative Certification	
Supervisor/Assistant	Not Valuable	Highly Valuable	Not Valuable	Highly Valuable	Not Valuable	Highly Valuable
Cooperating teachers	4.2	79.2	7.6	65.3	5.6	75.6
(N)		(24)		(118)		(90)
Principal	27.8	22.2	32.1	30.9	17.3	46.9
(N)		(18)		(84)		(98)
Other district personnel	43.8	25.0	40.7	25.4	14.8	36.4
(N)		(16)		(59)		(88)
University supervisors	4.5	59.1	16.1	33.1	14.3	71.4
(N)		(22)		(118)		(49)
Other university faculty	16.7	33.3	21.2	32.9	19.4	54.8
(N)		(12)		(85)		(31)
Other students in program	13.3	60.0	6.5	71.3	2.5	84.8
(N)		(15)		(108)		(79)

NOTE: Ratings based on a 1–5 scale, with "1" labeled "not at all valuable," and "5" labeled "extremely valuable." The "highly valuable" rating shown here includes both "4" and "5" responses.

The assistance and advice provided by fellow students is rated very highly. Between three-fifths to four-fifths of the recruits rated such advice as "highly valuable," suggesting that structuring programs to provide a cohort group with whom recruits can interact on a regular basis is very helpful to students. Many of these programs do, in fact, stress the value of the cohort experience (see Hudson et al., 1988).

Recruits' Perceptions of Program Strengths and Weaknesses

Participants and graduates gave remarkably similar responses when asked to list their programs' strengths and weaknesses; however, each type of program seems to have its own set of strengths and weaknesses, deriving from its unique nature and particular recruitment target pools. Recruits' opinions of program strengths and weaknesses are summarized in Table 5.5.

Program Strengths. Midcareer recruits rated the high quality of staff and fellow students as a program strength, and considered the teaching practicum and the relatively short, intense nature of the

Table 5.5

RECRUITS' EVALUATION OF PROGRAM STRENGTHS AND WEAKNESSES, BY PROGRAM TYPE

Program Strengths and Weaknesses	Mid-career	Recent B.A.	Alternative Certification	Retrainee
Strengths				
Good professors	(b)	(a)		(b)
Short, intense program	(b)	(b)		
Subject area courses		(b)		(b)
Teaching experience		(a)	(b)	
Teaching methods courses			(b)	
Classroom management			(b)	
Personal attention/tutoring				(b)
Paid internship	NA	(b)	NA	NA
Practicum	(a)			NA
Good students	(a)			
Cohort interaction			(a)	
Weaknesses				
Course content not useful	(a)	(a)	(b)	(a)
Too fast-paced/intense		(a)		(a)
Insufficient supervision		(a)		
Limited student teaching		(a)		

(a) At least 10 percent of total group mentioned item.
(b) At least 20 percent of total group mentioned item.

program to be positive program features. Here are some of the program strengths they listed:

> In a short period of time I was able to go into a classroom and teach with a fair amount of confidence.

> The program covered much ground in a relatively short period of time ... the professors did an excellent job with our group of eight candidates with varying backgrounds and ages.

> Brevity, M. Ed. degree option, and personalized attention were strengths.

Recent B.A. program recruits also found program staff and program intensity to be strengths, particularly training as a cohort and having the opportunity to enter the classroom quickly. In addition, these recruits praised their subject-area courses. The MESTEP recruits also considered their paid teaching and industry internships to be program strengths. Some examples:

> I found it most valuable in its placing of students in frequent contact with the others in the same program and the time to share ideas and concerns.
>
> Opportunity to independently handle a full class load for a semester and get *paid*; this, however, was also a weakness for me in that a lighter load would have been more beneficial and conducive to my creativity.... A strength was the individuals who comprised the program, especially the fellow "teachers to be."
>
> Excellent high school teachers taught methods course ... these instructors *modeled* excellent classroom teaching.
>
> Experience in industry and with computers was a strength, as was an internship in schools with full responsibilities.

Alternative certification recruits mentioned two course features—teaching methods and classroom management instruction—as program strengths, as well as opportunities to interact with other students in their cohort and the opportunity for immediate teaching experience. Their comments:

> The curriculum was excellent because it was extremely practical and applicable. Their visiting us on the job for a final evaluation was also very helpful. The intense but short time of our training was nice. The professors' encouragement and moral support was helpful, as well as the support group we formed—those of us going through the program.
>
> The program gets you into the classroom where you are needed and where you really learn to teach.
>
> The program was strong in preparing for instructional methods, particularly the frameworks for proper learning. Classroom management training was also good.
>
> The training program was extensive and thorough, with its greatest strength centering on interacting with experienced teachers and administrators.

Retrainees were most impressed by their subject-area courses, including course instructors and the personal attention and tutoring available:

> Coursework was presented in a very professional manner with much support from staff. Many opportunities were provided for tutoring. Coursework was very concise—a challenging experience.
>
> The teachers who worked with us were excellent. They made math exciting and were available for help.

> (The instructors) taught subject matter always with the idea we would be teaching this rather than using it for higher studies or going into the field.

In sum, these students show an appreciation for the kinds of program features that any student would desire—good instructors, relevant coursework, a supportive environment. But they also appreciate two program features that are more unique to these nontraditional programs: the short time frame for completing the program and the relatively quick exposure to the classroom.

Program Weaknesses. Some recruits felt that a major program weakness was that the course content was not useful or practical—that it ignored practical teaching issues, failed to link theory and practice, or did not focus enough on specific teaching methods for mathematics and science. Many recruits felt the need for more subject-specific pedagogical training, on the one hand, and more useful general instruction in classroom management issues, on the other. A sample of comments:

> Some of the coursework was required by the state for certification but was not too useful otherwise. (Midcareer recruit)

> Weaknesses fall in the area of teaching methods (methods to teach *specific* topics in one's subject area), linking theory with practice. (Recent B.A. recruit)

> No practical skills, such as discipline strategies, grading methods, "tricks of the trade" were offered. The program could use more experienced teachers as lecturers. (Recent B.A. recruit)

> Lack of student teaching, overemphasis on higher-level subject courses to the detriment of courses in classroom management, discipline, and learning styles, etc. (Recent B.A. recruit)

> Need more ideas for teaching particular units in a subject; need scenarios for discipline problems; education theory was virtually a waste of time. (Alternative certification recruit)

> Feel that more time should be spent in teaching the methods used to "teach" mathematics. (Retraining recruit)

Complaints about education coursework are not unusual among those who go through "traditional" programs (e.g., Applegate and Lasley, 1985; O'Rourke, 1983; Pigge, 1978; Ryan et al., 1979). Like these recruits, many teachers express a desire for coursework that is more directly related to teaching practice. The complaints offered here suggest that recruits are impatient with courses they view as irrelevant or too theoretical, but they favor increasing the pedagogical coursework that has practical utility for instruction in specific subject areas and with specific (hard to reach) students.

Though the short duration of most programs was listed by many as an attraction, it also often proved to be a problem for recruits. Alternative certification, recent B.A., and retraining recruits often felt that their programs were too fast-paced and intense, and that they were not fully prepared as a consequence. These comments are typical:

> Too much information in too short a time—not enough time to digest it all. (Alternative certification recruit)

> The program is too short (time wise) to absorb so much information. (Alternative certification recruit)

> Although I was very attracted to the short amount of time necessary to complete the program, it was a lot of knowledge packed into such a short time to absorb it all. (Retraining recruit)

> Time was extremely fast. Needed to slow down to give the students a chance to absorb course pedagogy. (Retraining recruit)

> A weakness was scheduling too much education coursework concurrent with the student teaching. People used to fall asleep on their desks at night after student teaching all day, maybe coaching, coming to class, and then staying up to prepare lessons. (Recent B.A. recruit)

> The program did not prepare me for the realities of the classroom. I needed more support during my internship—was not ready to be alone in the classroom. (Recent B.A. recruit)

> Not fully prepared for teaching internship; not enough *quality* student teaching time; not enough *practical* information. (Recent B.A. recruit)

RECRUITS' RECOMMENDATIONS FOR PROGRAM IMPROVEMENT

Recruits were asked about recommendations they might have regarding changes to the program in each of three areas: coursework, classroom teaching, and administrative services. Their responses were similar across program types, and addressed many of the program weaknesses recruits had identified.

Coursework

Regardless of respondent type, the most frequently mentioned need was for additional coursework in teaching methods, presenting varied instructional approaches and ideas on how to make science and mathematics courses more interesting and effective. The desire for more *specific* subject-matter pedagogy for particular courses and concepts was often mentioned. Those not in retraining programs recommended replacing or reshaping more theoretical education courses to make them more practical by focusing on the "realities" of classroom teaching. Alternative certification recruits wished they had had training in teaching methods and classroom management *before* they entered the classroom as full-fledged teachers. The retrainees, on the other hand, felt a need both for a greater number of subject-area courses and for a somewhat slower pace. Some felt that the courses should also be made more relevant by focusing on laboratory experiments or on material more applicable to the junior/senior high school levels, as opposed to the college level.

Program graduates also emphasized adding courses on classroom management, and on "practical" management skills, such as day-to-day planning of lessons, time management, class preparation, and grading. They also felt that experienced classroom teachers would be valuable as program lecturers or instructors.

Here is a sample of recruits' recommendations with respect to program coursework:

> Have outstanding teachers speak on how they have handled different situations rather than people with little or no classroom experience conducting seminars. (Alternative certification recruit)
>
> Would wish that we could have a methods course for our subject areas as part of our program in the summer before we start teaching. (Alternative certification recruit)
>
> Preparation for discipline *before* we went in the actual class would have been very helpful. (Alternative certification recruit)
>
> The coursework should be more practical than theoretical. Theories don't help when you're under fire in the classroom. (Recent B.A. recruit)
>
> Make coursework more relevant to actual teaching. Perhaps have seminars where experienced teachers come in and talk with students in the training program. (Recent B.A. recruit)
>
> Pedagogy courses should be more closely integrated with actual teaching practice. Some students seemed to need but did not get

> refreshed in specific areas of subject content—content weaknesses should be identified and addressed. (Midcareer recruit)
>
> I would require more hours in your field and less in education. All education classes should be taught with respect to your teaching field. For example, instead of a "teaching methods" class, I would rather see a "how to teach Algebra II effectively" class. (Retraining recruit)
>
> Require a methods course instead of one of the high level math courses. The highest level course material would rarely be needed to teach secondary math, whereas methods are always needed. (Retraining recruit)

Practicum

Most of the participants had just begun this aspect of their programs; apart from recommending a lengthier and more varied observation period and more supervision and feedback, comments were rather sparse. The one exception was the retrainee group, several of whom suggested that opportunities to observe mathematics and science teachers would be a useful addition to their retraining program.

Graduates had three major recommendations: more—and more varied—observation of experienced teachers, as well as of a variety of schools and students; more actual teaching practice; and closer supervision, together with more constructive and frequent feedback. This latter was the most frequently mentioned recommendation. Graduates of the North Carolina and South Carolina programs mentioned that using videotapes of their own teaching was extremely useful and that more of it should be done. (Some who had never seen their videotapes expressed interest in doing so.) The midcareer graduates felt that better selection and training of mentor/cooperating teachers and more interaction with these teachers would benefit the program. The South Carolina graduates, subject to a state-mandated evaluation program for licensure, mentioned the need for more than one observer during the practicum, for more objective and standardized evaluations and, in particular, the need for early feedback so that improvements could be made before their first year evaluation (which plays a major role in determining whether these recruits will remain in their program and in teaching).

Here are some comments addressing ways to improve the practicum experience:

> More! I would have liked to do some observation *after* my student teaching—it would have helped me see how other teachers did things—it means more once you've been through it. I would have

liked to have had more observation and feedback from supervisors. (Midcareer recruit)

Train cooperating teachers. Student teachers should talk with first-year teachers: They are closer in terms of their concerns and what is really necessary to survive. They share more similar experiences. (Midcareer recruit)

More time! Eight weeks was just not enough to get into the swing of it. (Midcareer recruit)

In my first year of teaching I have only been observed three times. I would like more feedback. (Alternative certification recruit)

There should be more informal observations earlier in our program, so that we may receive feedback and advice (only). (Alternative certification recruit)

I would have liked to have had a chance to observe experienced teachers in the classroom. (Alternative certification recruit)

Allowing us to see the videotapes of our peer teaching with individual instruction on areas to improve. (Alternative certification recruit)

More time during student teaching for *observation* of master teachers ... time for observing various teachers during internship. (Recent B.A. recruit)

Allow observation of several teachers in several schools to give an idea of different styles of teaching and different types of problems in the classroom. (Recent B.A. recruit)

Require more of a typical student practicum working with a wide variety of student ability levels and social backgrounds. (Recent B.A. recruit)

More observation and more thorough evaluations and critiques. Need better constructive criticism and evaluation vehicles. (Recent B.A. recruit)

The teaching internship was a wonderful experience, but would have been vastly improved by the addition of regular supervision. I couldn't believe that I taught one whole semester full-time and was observed only *twice*. (Recent B.A. recruit)

Administrative Services/Other

Frequently mentioned recommendations for improving administrative and other aspects of the preparation program included improved placement assistance for the practicum, and the provision of better information about requirements and certification status. The retrainees felt the scheduling of courses could be improved and also that better coordination between the program and state or county administrators and school principals was needed. Also, most of the program graduates felt the need for better placement assistance in obtaining teaching positions. The need for counseling or a mentor teacher, to assist in adjusting to the "real" teaching world, was stressed by several graduates. Many recruits also felt that more publicity to school districts about the program, with special focus on the certification status of the participants, was needed.

Some respondent comments:

> All of the candidates were confused about the job placement/hiring procedures, and the local school district personnel were just as vague in disclosing it. Some consideration should be given to this particular group of nontraditional recruits because of their age and practical experience. (Midcareer recruit)

> They should offer counseling to those who want or need it. Many people find their first classroom experience traumatic. (Recent B.A. recruit)

> Somehow the last leg of the recruitment/training process should be pairing with a good "mentor teacher" in the school hiring the new trainees. For me, that was probably the single most important thing affecting my development as a teacher. (Recent B.A. recruit)

> More advertising to school systems about who we were in order to make opportunities—jobs—open to us. (Recent B.A. recruit)

> Principals and district personnel should be required to be knowledgeable of the program. At present, only the participants know of the program. (Alternative certification recruit)

> Include a state list of openings for math and science teachers along with placement assistance and follow-up. (Alternative certification recruit)

> The scheduling of four classes back-to-back last summer was too intensive for most of us, I think. (Retraining recruit)

> Most of us felt that our counseling as far as credential requirements left much to be desired. There were many questions unanswered. (Retraining recruit)

Could have used assistance and counseling in securing interviews and placement for junior high position. (Retraining recruit)

SUMMARY

Judging by recruits' comments, these programs appear to be successful at meeting their basic goal: preparing nontraditional recruits to enter the classroom quickly. In so doing, however, the programs face the same conflicts as traditional teacher preparation programs, perhaps here accentuated by the need to prepare teachers in a shorter period of time. For example, the programs must provide sufficient coursework in teaching methods, balance theory and practice, and pace instruction appropriately. The programs seem to be somewhat uneven in their ability to address these issues: Some programs' recruits appear satisfied with their preparation in instructional methods, and others seem less so; the same is true for the teaching of classroom management, as well as the overall pace of the program. None of these appear to be severe shortcomings, although the emphasis in these programs on preparing teachers within an abbreviated timeline or curriculum may make resolving these issues more difficult. For example, the desire to have recruits enter the classroom quickly must be balanced by recruits' need to be adequately prepared for this experience, without being overwhelmed by an overly intensive course load. Also, the costs of providing an extensive and more closely supervised practicum must be weighed against recruits' desire for a short, affordable preparation program.

In fact, the nontraditional programs that follow a more "traditional" preparation approach—providing substantial pedagogical coursework *before* recruits enter the classroom and providing supervision and graduated assumption of responsibility during an internship—are more effective in the eyes of their participants and graduates. Programs that severely truncate coursework and place candidates on the job without adequate preparation or supervision are, not surprisingly, least well-rated by recruits.

Recruits' recommendations for program improvement focused on four ways of assisting recruits' adjustment to teaching: (1) making educational coursework more rigorous, more specific to subject matter pedagogical needs, and more practically informative; (2) providing longer, more varied and more closely supervised teaching practicum experiences; (3) providing better placement assistance for both the practicum and for those seeking teaching positions; and (4) providing mentor teachers or other assistance once in the classroom. The usefulness of involving expert, experienced classroom teachers as both course instructors and supervisors was also frequently noted.

Complaints that coursework is too theoretical, irrelevant, and impractical are fairly common among education majors. These kinds of complaints are common among students in other areas as well (e.g., Stolzenberg, Abowd and Giarrusso, 1986; Stolzenberg and Giarrusso, 1987). Interestingly, the utility of coursework and an understanding of the applications of theory seems to increase after students have spent some time in the classroom, suggesting the importance of *practice* to the understanding of *theory* and the importance of an iterative, staged process of preparation. This helps to explain, too, the importance and relevance that teachers place on their observation by peers. This is viewed as a valuable learning experience not only by new teachers, but also by experienced teachers, who find the observation of their colleagues to be a useful means of learning to apply their "old" teaching skills to a new subject area.

The ubiquitously expressed need for a longer or "better" practicum experience also reinforces the importance that traditionally trained teachers place on this aspect of teacher preparation (e.g., Ryan et al., 1979). Among these nontraditional recruits, however, the two groups that do not have a student teaching/practicum experience (retrainees and alternative certification recruits who enter the classroom without prior practice teaching) also stressed the need for one. Those in retraining programs felt that observation of mathematics or science teachers would assist them in learning to apply their new knowledge, whereas alternative certification recruits felt that observation before entering the classroom would improve their initial teaching performance and adjustment. The alternative certification recruits keenly felt the need for more and better supervision on the job. Shockingly, given the emphasis these programs place on on-the-job training *in lieu* of much generally required coursework, they in fact offered the least supervision of any program type, with many recruits receiving assistance only a few times during an entire year.

The responses discussed here provide valuable insights for those interested in developing or improving programs to best meet the needs of the beginning teacher. For example, suggestions concerning the pacing and level of courses, increased opportunities for classroom observation, more varied or typical field placements, and better supervision of these placements should all be given serious consideration. However, it is important to note that these suggested changes are also likely to increase program costs and length.

VI. NEW RECRUITS' TEACHING EXPERIENCE AND FUTURE PLANS

An important measure of the success of nontraditional recruitment policies is the degree to which program graduates enter and remain in teaching. We examine these issues in this section, by assessing program graduates' current occupational status, satisfaction with teaching, and future plans. Since members of different segments of the reserve pool come to teaching at different points in their lives and from various careers, they may vary in their expectations of and commitments to teaching. Consequently, data in this section are disaggregated based on respondents' major activity in the year before program entry—i.e, K-12 teaching, student, working in science or engineering fields, and "other."[1] For convenience, we will refer to these recruit types as: former teachers, former students, former science workers, and others.

TEACHING STATUS OF GRADUATES

Table 6.1 presents the current teaching status of program graduates.[2] About three-quarters of those who had not previously been teachers were in K-12 teaching positions at the time of the survey. For continuing teachers, the figure was just over 90 percent. Former students were less likely than other recruit types to enter teaching after program completion; this is not surprising, as these younger recruits may have a broader range of opportunities open to them. Not having invested in a major career shift, they might be less committed to teaching as a career pursuit. Nonetheless, the proportion of graduates who entered and remained in teaching—at least initially—is fairly high. Excluding former teachers, 86 percent of all nontraditional program graduates entered teaching, and 75 percent were still in teaching within an average of two years of program completion. This compares quite favorably with data on other recent college graduates who prepared for teaching. For example, the 1985 Recent College Graduates Survey found that only 60 percent of bachelor's degree recipients

---

[1]See Table 4.2 for the distribution of these recruit types among programs.

[2]These data show entry rates into teaching for different recruit types; however, they cannot be interpreted as retention rates because of the variation among recruits in length of time since graduation. For example, most graduates had completed their program two years ago, but the range spans from less than one year to six years.

Table 6.1

CURRENT TEACHING STATUS OF PROGRAM GRADUATES
(In percent)

	Currently Teaching K–12	Not Now Teaching K–12[a]	Never Taught K–12	(N)
Recruit type				
Former teachers	91.4	7.5	NA	(93)
Former students	70.4	11.1	18.5	(54)
Former science workers	74.4	12.8	12.8	(39)
Other	81.8	9.1	9.1	(44)
Program type				
Midcareer	71.4	7.1	21.4	(14)[c]
Recent B.A.	70.0	16.0	14.0	(100)
Alternative certification[b]	96.5	0.0	3.5	(29)[c]
Retraining	91.4	6.7	1.9	(104)

[a] Have taught at the K–12 level since completing teacher preparation program but not currently teaching at this level.

[b] In this table, we include only those alternative certification program students who have actually completed and graduated from their program. In other analyses of teaching experiences we include current alternative certification program "participants," as their program participation requires that they teach full-time.

[c] Small sample sizes warrant caution in generalization.

who were newly qualified to teach in 1983–84 were teaching one year after graduation. The higher rate in our sample may be due to greater job availability for mathematics and science teachers, greater likelihoods of seeking teaching jobs, or both.

Of the 19 recruits who did not enter teaching, 17 responded to the survey item that asked why they chose not to enter teaching. Although the sample size for this item is too small to allow for any definitive conclusions, recruits' responses (listed in Table 6.2) suggest that the desire to pursue another career, and dissatisfaction with teaching during the practicum, are major reasons for this failure to enter teaching. Only two recruits reported being unable to find a teaching job.

Table 6.2

RECRUITS' REASONS FOR NOT TEACHING AFTER GRADUATING FROM TEACHER PREPARATION PROGRAM

Reason	Percent[a]
Wanted to pursue another career	52.9
Did not enjoy teaching during student teaching/internship	41.2
Went on to further graduate or professional training	17.6
Chose to stay home and care for family	11.8
Could not find a teaching job	11.8
Other[b]	35.3
(N)	(17)

[a] Column total sums to greater than 100 percent because of multiple responses.
[b] Reasons given included failure to pass the NTE, dissatisfaction with the offered position, and conflict with philosophy of the educational system.

RECRUITS' APPLICATION DECISIONS

Among those who applied for teaching positions, recruits were asked about the factors that were important in their decision to apply to particular school districts. Table 6.3 shows that, for all recruit types, working conditions in the school and the school's location were the two

Table 6.3

IMPORTANCE OF CRITERIA IN CHOOSING SCHOOLS TO WHICH TO APPLY FOR TEACHING POSITIONS, BY RECRUIT TYPE[a]
(In percent)

Criteria	Former Teachers	Former Students	Former Science Workers	Other
Working conditions	71.8	60.7	67.4	71.9
Location	82.5	51.7	70.6	70.3
Salary	51.3	33.9	47.9	43.6
Academic reputation	42.5	33.3	51.0	37.7
Availability of supplies	44.7	33.9	36.7	31.7
Size of school	31.6	25.0	12.5	26.2
(N)	(40)	(58)	(51)	(64)

[a] Percentages shown are those rating the item as "very important" (a rating of three on a three-point scale).

major criteria used for application decisions, although both factors were slightly less important for former students than for other recruit types. Former students also placed less importance on salary than did other recruits, whereas teachers placed more emphasis on the availability of supplies and science-field workers placed more importance on the school's academic reputation. These differences suggest that: (1) Although students may be less likely to enter teaching, they are more flexible than other recruits about the type of school to which they apply; their youthfulness may make them more willing to deal with less ideal working conditions, salaries, or locations; (2) the availability of supplies and other materials for teaching, although not one of the most important criteria, appears to be an indication of favorable working conditions that is recognized more by those who have already taught than by those without teaching experience; and (3) those who have been employed in science fields are more likely to want to teach in schools where academics, including the sciences, appear to be highly valued.

Recruits were also asked if they thought they had been perceived differently by school districts because of the type of program they attended. As Table 6.4 shows, graduates of recent B.A. programs were especially likely to believe that they were viewed differently from other job applicants. Only one-half to just over one-third of graduates of other program types felt that they had been perceived differently. The verbatim responses to this question showed that recruits' negative or positive perceptions varied by program type. Positive biases appeared to arise because of the "good reputation" of some programs and of their graduates' knowledge; negative reactions stemmed from skepticism about the nature of the recruits' preparation in these nontraditional programs and from "resentment" on the part of other educators toward those who enter the system through alternative certification routes.

Table 6.4

PERCENTAGE OF RECRUITS WHO BELIEVED THEY
WERE PERCEIVED DIFFERENTLY BECAUSE OF
THE TYPE OF PROGRAM THEY ATTENDED

Program Type	Percent
Recent B.A.	76.5
Midcareer	53.8
Retraining	39.3
Alternative certification	37.9

For example, graduates of recent B.A. programs appeared to be very favorably received, primarily because of their programs' good reputation and the successful teaching experiences of former graduates. Only five of the midcareer program graduates felt they had been viewed differently on the job market. Three of these felt their ability to teach their subject area was perceived as a strength, although two felt their qualifications were viewed skeptically. Most of the retrainees did not apply for a job after program completion, primarily because they were already employed (and often in the subject-area for which they were being retrained). Of those who did apply for new positions, most felt they were not viewed differently or were viewed positively because of their new knowledge and credential. Although most graduates of alternative certification programs did not believe they had been perceived differently, those who did feel they were treated differently encountered many more negative than positive reactions. Skepticism of their teaching ability and their certification status and resentment from other teachers were frequently mentioned problems.

CERTIFICATION STATUS AND TEACHING POSITION

To characterize the types of teaching positions filled by these nontraditional program graduates, we asked them about their areas of certification, and, for those who taught at any time after graduation, about their teaching assignment, including the type of school in which they are employed and the types of students they teach.

Certification Status

Ninety-six percent of all graduates listed at least one subject area in which they are currently certified; the distribution of these responses is provided in Table 6.5. As this table shows, nearly all retrainees are being certified in mathematics, general science, and the biological sciences; other program graduates become certified in a broader range of areas, with much greater representation in the physical science fields. A substantial number of new recruits become certified in two or more areas of science and mathematics.

Main Teaching Assignment

Graduates' main areas of assignment are listed in Table 6.6. There are two interesting findings in this table. First, 40 percent of retraining program graduates have a main assignment area *outside* of

Table 6.5

GRADUATES' AREAS OF CURRENT CERTIFICATION
(In percent)

Areas of Current Certification	Retraining Program Graduates	All Other Graduates
General science	14.1	41.9
Earth/space science	1.9	28.4
Biological/life sciences	19.8	42.4
Physical sciences	4.7	42.4
Mathematics/computer science	48.1	35.2
Elementary education	34.9	0.4
English/language arts/reading	17.9	0.4
Physical education/health	13.2	0.0
Social studies	16.0	0.8
Other	43.4	5.1
(N)	(106)	(236)

Table 6.6

GRADUATES' CURRENT MAIN ASSIGNMENT AREA
(In percent)

Main Assignment Area	Retraining Program Graduates	All Other Graduates
Biological/life sciences	7.2	24.2
Chemistry	0.0	12.1
General science	2.4	9.5
Earth/space science	2.4	8.9
Physics	0.0	6.8
Algebra	9.6	12.1
General mathematics	21.7	7.9
Geometry/trigonometry	2.4	4.2
Computer science	6.0	2.6
Remedial mathematics	6.0	2.6
Other mathematics[a]	0.0	1.1
Other areas	19.3	6.8
Self-contained class	20.5	1.1
(N)	(83)	(190)

[a]Includes the following categories: business/consumer mathematics, probability/statistics, and calculus.

mathematics and science, suggesting that many of these graduates have not yet had an opportunity to use their retraining. Second, relative to their certification fields, few graduates are assigned mainly in the physical sciences; this suggests that teachers certified in these areas are also providing a significant amount of instruction in other areas. Given the magnitude of reported teacher shortages for chemistry and physics, it is surprising that fewer than 20 percent of new recruits are primarily assigned to teach in these areas. However, in most high schools, only a small fraction of students elect—or are selected—to take physical science courses; many more are required to take biology and other life science courses. Consequently, teachers trained in the physical sciences must often also teach life sciences, mathematics, or other courses as part of a full teaching load.

Despite the multiple certifications obtained by many new recruits, a fair number are assigned to teach primarily in a field for which they are not certified. Comparing graduates' main assignment areas with their fields of certification shows that, overall, 82 percent of all recruits with a main assignment area in mathematics or a science field are certified in that field. Thus, by this measure, 18 percent of these new teachers are misassigned in their main area of assignment. As Table 6.7 shows, this misassignment is most common in earth science, and least common in the biological/life sciences. These data do not reflect additional out-of-field assignments for courses outside recruits' primary assignment areas; hence they underestimate the extent to which recruits teach one or more courses outside their certification fields.

Table 6.7

PERCENTAGE OF GRADUATES CERTIFIED
IN MAIN ASSIGNMENT AREA

Main Assignment Area	Percent
General science	80
Earth science	68
Biological/life sciences	87
Physical sciences	81
Mathematics/computer science	83

Teaching Position

To learn more about the types of teaching positions in which recruits are placed, graduates were asked for the following information about their school and students: school locale, school sector, grade levels taught, and students' socioeconomic status (SES) and achievement level. As Table 6.8 shows, recruits work almost exclusively in the

Table 6.8

SELECTED CHARACTERISTICS OF GRADUATES' SCHOOLS
AND STUDENTS, BY RECRUIT TYPE
(In percent)

Selected Characteristics	Former Teachers	Former Students	Former Science Workers	Other
Type of school				
Private religious	0.0	8.9	1.8	2.4
Private nonreligious	1.8	5.4	0.0	2.4
Public	98.2	85.7	98.2	98.2
(N)	(112)	(56)	(57)	(83)
Community type				
Rural	10.7	21.4	12.1	18.5
Small town	15.2	25.0	24.1	29.6
Suburb	41.1	28.6	20.7	35.8
Urban	33.0	25.0	43.1	35.8
(N)	(112)	(56)	(58)	(81)
SES of schools' student body				
Lower or lower-middle	42.5	31.5	34.5	49.4
Middle	20.4	14.8	24.1	17.3
Upper-middle or upper	20.4	18.5	13.8	7.4
Wide range	16.8	35.2	27.6	25.9
(N)	(113)	(54)	(58)	(81)
Student achievement level				
Primarily high	14.2	17.9	19.0	21.7
Primarily average	38.9	23.2	27.6	20.5
Primarily low	19.5	21.4	22.4	33.7
Widely differing	28.3	37.5	31.0	24.1
(N)	(113)	(56)	(58)	(83)
Grade level				
Elementary (K–6)	20.4	3.6	8.6	7.3
Middle (7–8)	37.2	23.2	22.4	29.3
Senior high (9–12)	56.6	91.1	79.3	75.6
(N)	(113)	(56)	(58)	(82)

upper grades of public schools, across a wide range of locales, school SES levels, and student achievement levels. As the assignment data implied, about 20 percent of former teachers (mainly retrainees) are teaching at the elementary level; these teachers have not yet found teaching assignments appropriate to their new training.

Compared to all teachers, these recruits are more likely to be employed in urban school districts (NEA, 1987)[3] and appear to be disproportionately placed in schools serving low-income students. This coincides with other research showing that beginning teachers tend to be placed in schools with the highest turnover rates, frequently those serving more disadvantaged students (Wise, Darling-Hammond, and Berry, 1987). Beginning teachers also tend to be given the least desirable teaching assignments. This may help explain some of the dissatisfaction and resentment expressed by recruits regarding their teaching assignments; these reactions are discussed below.

RECRUITS' SATISFACTION AND ADJUSTMENT TO TEACHING

Graduates who had taught following program completion were asked about their satisfaction with and adjustment to various aspects of their teaching experience, including their current teaching assignment and the school workplace. Graduates were also asked whether they have ever been so dissatisfied with teaching that they had considered leaving the profession, and, if so, what was the most important cause of this dissatisfaction.

Teaching Assignment

Most graduates are basically satisfied with their teaching assignment; however, most also believe that their assignment could be improved (Table 6.9). Graduates' preferences are remarkably consistent; for all recruit types, the most strongly desired changes in assignment are fewer or smaller classes, and more courses at a higher level of instruction. Write-in responses revealed other dissatisfactions that are not necessarily assignment-related: desire for more motivated students, less paperwork, fewer nonteaching duties, more time for preparation, better equipment and, among former teachers, assignments in the area for which they were retrained. However, by far the most frequent write-in response was a desire for a higher salary. These

---

[3] NEA data show that only about 22 percent of all teachers are employed in urban schools; the proportions in our sample range from 25 to 43 percent (NEA, 1987).

Table 6.9

GRADUATES' SATISFACTION WITH THEIR TEACHING ASSIGNMENT
AND DESIRED CHANGES IN ASSIGNMENT, BY RECRUIT TYPE
(In percent)

Characteristics of Teaching Assignment	Former Teachers	Former Students	Former Science Workers	Other
Satisfaction with assignment				
Unsatisfied	9.7	14.6	12.1	25.3
Satisfied	50.4	49.1	56.9	32.5
Very satisfied	39.8	36.4	31.0	42.2
Those stating that some change would make assignment more satisfactory	77.7	85.7	87.7	73.5
Desired change in assignment				
Fewer or smaller classes	64.4	64.6	70.0	52.5
More courses at a higher level of instruction	46.0	51.8	44.0	52.5
Different grades	19.5	14.6	10.0	16.4
Fewer courses outside area of expertise	10.3	8.3	14.0	9.8
More varied course topics	10.3	12.5	12.0	6.6
Other[a]	23.0	33.3	30.0	29.5
(N)	(113)	(56)	(58)	(83)

[a] See discussion of open-ended responses in text.

complaints are fairly typical of those given by all teachers. Altogether, these data suggest that the new recruits face fairly typical problems in their new teaching assignments.

Teaching Experience

There was a high level of consistency in the satisfaction levels reported by different recruit types with various aspects of teaching (Table 6.10). Despite the different experiences recruits brought with them into teaching, they ranked aspects of their new occupation very similarly. Overall, the majority of graduates were satisfied with the following: flexibility in deciding how to teach, success with students, peer relationships, administrative support, teaching opportunities, nonteaching duties, and available resources. Only about half are satisfied with their chances for professional advancement, student respect for

Table 6.10

PERCENTAGE OF GRADUATES SATISFIED OR VERY SATISFIED
WITH VARIOUS ASPECTS OF TEACHING, BY RECRUIT TYPE
(In percent)

Teaching Characteristic	Former Teachers	Former Students	Former Science Workers	Other
Flexibility deciding how to teach	94.8	98.2	94.7	89.2
Success with students	92.2	83.6	81.0	84.2
Collegial relationships	89.6	91.1	75.4	82.9
Support from administrators	69.6	66.1	67.2	73.2
Opportunities to teach classes you like	69.3	76.4	63.8	66.3
Nonteaching duties	63.5	55.4	55.2	69.9
Resources and materials	59.1	53.6	53.5	59.8
Chances for professional advancement	49.6	43.6	53.4	63.9
Student respect for teachers	54.8	55.4	44.8	50.6
Class size/teaching load	47.8	55.4	44.8	56.1
Support from parents	46.1	47.3	46.6	52.4
Discipline in the school	57.4	39.3	41.4	44.6
Salary	39.1	26.8	39.7	45.8

teachers, their teaching load, and parental support. Finally, fewer than half of all graduates were satisfied with school discipline, and, at the bottom of the list, salary.

These patterns follow those among teachers in general. For example, a 1984 Gallup poll found that secondary school teachers were most dissatisfied with the following educational conditions: discipline, parental support, financial support, pupils' lack of interest, and problems with administration (Gallup, 1984). This survey did not ask specifically about teacher salaries or teaching loads, but a National Education Association survey (1987) found that "heavy workload/extra responsibilities" was rated by middle and high school teachers as a major occupational problem, as was "lack of funds/decent salaries." This survey also found teachers to be dissatisfied with student discipline and administrative support. Another recent survey examined teachers' actual versus ideal class size, and found that middle and high school teachers want, on average, a 16 percent reduction in class size (from 25 to 21 students) (Metropolitan Life, 1986). Interestingly, although both the Gallup poll and NEA survey found teachers to be dissatisfied with administrative support, most of the teachers in this study were satisfied with this aspect of teaching.

Table 6.10 does show a few interesting differences among recruit types. First, former teachers are *more* satisfied than other recruit types with their success with students and with school discipline. This is probably related to their greater experience—former teachers have had the chance to learn how best to succeed with students (or to leave teaching if they were not successful), and they are probably more adept at handling discipline problems. Generally, however, former teachers and other new recruits have very similar perceptions—reinforcing the finding that "traditional" and "nontraditional" entrants view teaching quite similarly.

Former students and former science workers also have slightly different reactions to some aspects of teaching. Students are more likely to be satisfied with their opportunities for teaching classes that they like; this may reflect either their greater willingness to teach a variety of courses, or their better preparation for teaching a wider range of courses. Students are also more dissatisfied with their salary; perhaps they enter the teaching profession with somewhat unrealistic expectations, unlike older individuals with more experience in the labor market. (Some older recruits may also have less pressing financial needs, if they are receiving some retirement benefits, have already purchased a home, or have received credit for their greater work experience and education in their teaching salaries.) Finally, former scientists tend to be *less* satisfied than other recruits with their peer relationships and with the respect they receive from students. This may reflect the difference between the perhaps more collegial structure for conducting scientific work, compared to the more isolated relations among teachers (Rosenholtz et al., 1985) and the sometimes antagonistic relations between students and teachers.

Expectations of Recruits

A final question asked recruits how their on-the-job teaching experience differed from their initial expectations or from their student teaching experience. Responses to this question suggest that recruits were not fully prepared for two major aspects of public school teaching: the need to motivate students, and the demands of teaching and non-teaching duties on time and energy. More specifically, recruits did not expect to encounter the following student characteristics as frequently as they did: lack of respect for teachers, discipline problems, lack of motivation, and low achievement levels. They also complained of apathetic parents. In a few cases, recruits specifically mentioned that the students they are currently teaching are less motivated than those with whom they student taught.[4] Their comments also point to the need for more and better training regarding student motivation and behavior. Here are some of their comments:

> Students in my school lack motivation and ambition. They really don't want to be in school and don't appreciate your attempts to help them. It's hard beating your head against the wall.

> The biggest difference was in how much I had to work to make teaching and learning interesting—how much I had to bring to my subject beyond the basic facts and processes.

> The students I taught as a student teacher were of a special breed—they attended a high school in a major university town . . . . My students were very different. Most of them couldn't care less and were very unmotivated.

> The students sometimes differ drastically from the "model" students spoken of in class. Assertive discipline, taught in the program, works only with certain students in certain schools. Student respect was much lower than I expected. Also, students' emotional and psychological problems are more common.

> Discipline has been a difficult issue for me. I am now teaching remedial math to high school students. I never expected to have to teach arithmetic to high school students or to have to deal with the kinds of behavior problems these kids have. I have quickly learned to be stricter with these students, but I had to change my expectations.

> My expectations regarding students' motivational levels were somewhat shattered. Student and parent apathy is something I've had trouble dealing with. Discipline, too, is of deep concern to me.

These graduates also often found that teaching involves more time and energy than they had expected, especially for paperwork and "mindless duties." They complained of having too little time for preparation or for their students and of bureaucratic deflections from teaching. They also described teaching as "draining":

> I had no idea how demanding actual instruction and preparation would be! Teaching is extremely hard work! Knowledge of subject matter is only one small ingredient in successful teaching.

---

[4]This may be partly due to differences in the types of schools and classrooms to which student teachers are assigned and those where beginning teachers are generally placed. It may, however, not be entirely due to objective differences in students; new teachers probably do not know as much about motivating students as their cooperating teachers did.

> Teaching is too paperwork-loaded. Little time is available for preparation or teaching or even for evaluating students. Most time is spent conforming to bureaucratic duties.

> I never expected to be so exhausted at the end of the day; I didn't expect so many extra jobs like lunch duty ... didn't expect such a lack of understanding of the part of the administration concerning the importance of class time (constant interruptions for pep rallies, assemblies, etc.).

> The amount of time required for nonteaching, record-keeping duties is enormous. My lesson plans are the last thing I do, and I have little time for them .... I feel the system prevents me from making the decisions and providing the opportunities I need to provide for my students.

A final complaint mentioned by a number of teachers was that new teachers seem to be given the least desirable assignments—a situation that they neither expected nor viewed as professionally sound. Their comments:

> I thought that there would be higher level courses for me to teach and instead when I arrived at the school I found that all of the lower level courses were given to me.

> I did not expect to be given so many remedial algebra classes my first year (four).

> As a teacher, one expects to be treated as a professional .... This has not happened to me. Teaching is the only professional job I know where they give you the crappiest and hardest job (i.e., low-ability kids, large class size, changing rooms between periods) to those with the least experience to handle it. Three other people in my 12-person accelerated certification program have quit for similar reasons. The principals ... give the good assignments to the rockstars of the high school world (i.e., coaches).

These complaints were fairly pervasive; however, a large minority described teaching as more rewarding than they had expected. They often mentioned the strong positive relationships they had with their students, or stated that having full responsibility and control in the classroom was an improvement over student teaching. Many of these respondents mentioned both the positive and negative ways in which teaching differed from their expectations. Here are some examples:

> I have greater enjoyment in having the "helm." I enjoyed my student teaching and my extensive substitute assignments, but actually watching student growth/achievement from the year's beginning to end has meant *very* much to me.

Having student taught prior to my training experience, I did not experience any big surprises. The volume of nightly work and preparation was a difference (an increase), but the independence of having my *own* class was a pleasant difference.

It is much harder work, but much more fulfilling than I had anticipated.

The actual (teaching experience) was much more relaxing. They were *my* kids and I wasn't a pawn.

Students were quite receptive, and it was easy to *earn* respect, and establish rapport.

Dissatisfactions

When asked if they had ever been so dissatisfied with teaching that they had considered leaving the field, about one-quarter to one-third of these new recruits replied "yes" (Table 6.11). This is a lower percentage than that for teachers in general. For example, a Metropolitan Life survey (1986) found that 55 percent of all teachers had considered leaving teaching. Recent NEA surveys have found 30 to 40 percent probably would not enter teaching if they had it to do over again (NEA, 1987). The lower proportion in our sample may be partly attributable to self-selection: Most of these recruits have, after all, made a conscious choice to transfer into teaching from other occupations and, therefore, may have a more considered commitment to teaching. The new recruits also have only a few years of teaching experience, whereas the average teacher has had 15 years in which to experience doubts or dissatisfaction.

When asked to give their major reasons for considering leaving teaching, recruits most often mentioned salary, school discipline, and (lack of) student respect. Former teachers also mentioned the lack of administrative support. These are typical findings. For example, the effects of salary on teacher retention are well-documented (Murnane and Olsen, 1988; Zabalza, 1979). Also, a recent study examining school discipline found that this factor alone had led 29 percent of all teachers to seriously consider leaving the field (CES, 1987b).

Table 6.11

PERCENTAGE OF GRADUATES WHO CONSIDERED LEAVING
TEACHING AND THEIR REASONS, BY RECRUIT TYPE

	Former Teachers	Former Students	Former Science Workers	Other
Percentage who have considered leaving	29.3	34.8	23.8	24.2
(N)	(116)	(66)	(63)	(91)
Major reasons for considering leaving				
Salary	*	*	*	*
Discipline		*	*	*
Student respect		*		
Administrative support	*			
(N)	(34)	(23)	(15)	(22)

FUTURE PLANS

Given the mixed evidence on the satisfaction derived from teaching, the question of how many of these new recruits plan to remain in teaching becomes an important one. This question is also a more practical concern, as the costs of nontraditional teacher preparation programs (which are borne mostly by third parties) are in some sense justifiable only if they can train qualified teachers to enter *and* remain in the classroom. It does the educational system little good to invest in special programs to prepare individuals with subject matter knowledge for entering teaching if these individuals then fail to enter teaching or to remain in the field for an extended period. We therefore asked program participants and graduates[5] what their future plans are with respect to teaching (Table 6.12).

Table 6.12 shows that few recruits (overall, 8.2 percent of the 86.8 percent that have not yet left) are planning to leave teaching immediately. An additional 15 to 24 percent are unsure of whether they wish to remain in teaching. Only about half (51.9 percent) of these recruits plan to make teaching their career, with former teachers and science workers showing a much greater inclination than others to become (or

---

[5] Separate analyses of responses given by graduates and participants showed no major differences, thus responses for the two groups were combined. This similarity of responses suggests that, in spite of its unexpected drawbacks, the actual teaching experience may have little effect on recruits' long-term teaching plans.

Table 6.12

RECRUITS' FUTURE PLANS WITH RESPECT TO TEACHING, BY RECRUIT TYPE
(In percent)

Future Plans	Former Teachers	Former Students	Former Science Workers	Other
Plan to make teaching my career	58.8	36.8	67.6	38.5
Plan to teach until moving into an administrative position	11.0	13.2	2.9	16.7
Plan to eventually pursue another career	9.3	13.2	4.4	9.4
Plan to leave teaching as soon as I can	4.9	14.7	5.9	11.5
Undecided at this point	15.4	22.1	19.1	24.0
(N)	(182)	(68)	(68)	(96)

[a]Graduates who did not enter teaching or who are no longer teaching are not included. Alternative certification program participants who are teaching full-time are included, though they have not "graduated" from their three-year program.

remain) career teachers. These groups include the more mature recruits who are less likely to envision further career switching.

Fewer than 40 percent of former students and "other" entrants (nonscience workers) plan to make teaching a career. Just over 70 percent plan to remain in teaching for at least a while (ranging from a low of 63 percent for former students, to a high of 79 percent for former teachers). These figures are roughly comparable to those of middle/secondary teachers in general, for whom about 6 percent plan to leave teaching as soon as possible, 20 percent are undecided, and about 75 percent plan to remain in teaching for at least a while (NEA, 1987).[6] Few recruits aspire to become administrators; former science workers are least likely to want to leave teaching for administration.

---

[6]When evaluating these results, it is important to keep in mind that the nontraditional recruits (with the exception of former teachers) are either in the early years of their teaching careers, or have not yet begun teaching; there is evidence to show that teachers in the first five years of teaching tend to have high attrition rates (Grissmer and Kirby, 1987). We do not know the extent to which these intentions data may accurately approximate the later actual attrition rate.

SUMMARY

Graduates from nontraditional programs appear to enter and remain in teaching at rates that are comparable to, and probably higher than, those for traditionally prepared teachers. Excluding those who were already teachers before program entry, 86 percent of program graduates enter teaching and about 75 percent are still teaching within (an average of) two years of program completion.

Graduates of these nontraditional programs do not have difficulty finding teaching jobs. In our sample, graduates of the recent B.A. programs, in particular, felt that they were very well-received by local districts. Among the programs we examined, familiarity with the program or its graduates seemed to encourage the acceptance of new program graduates. On the other hand, for some candidates, initial acceptance was difficult; this seems to be especially true for alternative certification program graduates, who felt that many school personnel were reluctant to accept an alternative certification program as a valid or thorough means of teacher preparation.

When asked about their satisfaction with teaching and their current teaching assignment, these nontraditional program graduates voice concerns very similar to those expressed by teachers in general, and by new teachers in particular. Like other beginning teachers, these new recruits experience some degree of "reality shock" (cf. Brown and Williams, 1977; Gaede, 1978), and find that student discipline and motivation are two of their most difficult problems (cf. Elias et al., 1980; Marso and Pigge, 1987; Veenman, 1984). They are also unhappy with their salary and with the respect they receive from students. Although it is reassuring to find that many new recruits find teaching to be more fulfilling and rewarding than they had expected, the preponderance of unexpected problems and disappointments suggest that, as others have noted, teachers need a better "reality base" from which to begin their teaching careers (Chapman, 1983), and more support within their first few years of teaching (McLaughlin et al., 1986; Wise et al., 1987).

The question of whether and how long these new recruits plan to stay in teaching is an important one. After all, nontraditional teacher preparation programs are in some sense effective only if they train qualified teachers to enter *and* remain in the classroom. Among our entire sample of participants and graduates, approximately 70 percent plan to remain in teaching for "a while" although only about half plan to make teaching a career. Recruits who were students just before program entry appear to be least likely to enter and remain in teaching. These individuals tend to have lower costs of retraining for another occupation; in addition, they may have a wider set of job opportunities available to them.

These data are roughly comparable to those for teachers in general. On this measure, these programs appear to be at least as successful as more traditional programs in preparing teachers for the classroom. However, given the later career choice that many of these new recruits have made, and the time and energy that many have devoted to training for this new career, one might expect an even higher level of planned retention. It seems, however, that it is the difficult nature of teaching itself that is largely responsible for most new recruits' considerations of leaving teaching. Working conditions feature heavily in expressed dissatisfaction, especially schools' apparent emphasis on paperwork, and nonteaching activities at the expense of teaching time. In one respect, though, program preparation may be a factor: Many recruits felt they could have been better prepared to handle classroom management and student discipline than they were. Though actual experience is necessary in developing this type of skill, better preservice training could ease the transition.

VII. CONCLUSIONS AND RECOMMENDATIONS

This study assessed the role of nontraditional teacher preparation programs in addressing current shortages of qualified mathematics and science teachers. We examined programs and their recruits to describe (1) the effective reserve pool for mathematics and science teachers; (2) the design and implementation of nontraditional teacher preparation programs; and (3) the characteristics and experiences of program participants and graduates.

We discovered that programs' successes at recruiting and preparing qualified entrants to teaching depend in part on how they design their educational programs, whom they try to recruit, and local labor market conditions. This section suggests strategies for overcoming some of the problems many programs have encountered and discusses conditions for continued success in tapping nontraditional recruitment pools for mathematics and science teaching.

PROGRAM VIABILITY

Although current nontraditional teacher preparation programs are quite diverse, they typically share a number of common features. In accordance with their goal of providing a more feasible alternative to traditional undergraduate programs, these programs tend to target coursework more closely on recruits' certification needs and are less costly in time and money than traditional programs. Focusing on one particular type of recruit—retirees, midcareer transfers, or former teachers, for example—allows a program to create a curriculum tailored specifically to the needs and prior training of that group.

Financial costs to recruits tend to be low, or in some cases, nonexistent, as many of the costs of the programs are covered by third-party sponsors—states, school districts, federal government or foundation grants, or industry donations. These total program costs can be substantial, ranging from approximately $2,500 to $10,000 per recruit. Procedures and arrangements that ease the financial and time burdens of teacher preparation are, not surprisingly, the program features that are most attractive to nontraditional recruits.

However, some of the same features that make programs attractive to potential recruits also make them vulnerable. Dependence on outside donors means that the programs may disappear when funds are scarce. Also, funding sometimes disappears when there is a perception that shortages have been "solved" or when other budgetary priorities take precedence for local or state agencies. As a consequence, eight of the 64 programs we initially surveyed had been discontinued by the time the study was completed. Several others were unsure that they would continue operating in the following year, and many others had substantially changed their focus, scope, or approach to preparation and recruitment.

Similarly, the narrow targeting of recruitment pools that allows programs to tailor their approach for a specific type of recruit can backfire if the target pool shrinks or is not responsive to the incentives these programs offer. Unless programs are flexible enough to expand their candidate pools, they may go out of business when local demographics or labor markets change. For example, when unemployment in science fields outside of education is relatively high (as in the Southwest when oil industry contraction occurred recently), midcareer recruits may be more easily found. When the industrial situation changes, other sources of recruits will be needed. Similarly, when early retirement incentives in industry are common, retirees interested in teaching are more available. As policies shift, these recruits are less numerous.

Nontraditional programs should be prepared for expansion and contraction with shifts in local teacher demand and supply, as program support often depends on there being a perceived shortage of traditionally trained teachers. This occurs in part because some types of programs tend to be viewed more as a means of alleviating local shortages than as a preferred or "legitimate" means of preparing teachers. In addition, since a substantial share of program costs is assumed by third parties, program funding may rely on the extent to which funders perceive a serious need for teachers.

In our study, retraining and alternative certification programs were most apt to be viewed as stop-gap measures for addressing teacher shortages. In retraining programs, this is because of the costs of training and reassigning current teachers. In alternative certification programs, it is because the reduced pedagogical coursework typical of these programs is often perceived as resulting in substandard preparation. Both types of programs also face the difficult task of providing adequate training over a brief period of time to individuals who are simultaneously working as full-time teachers—typically a 50-hour per week job before extra responsibilities are added. Intensive coursework added to this schedule creates a highly stressful situation, especially for those who are just beginning their teaching careers.

Some programs—especially university-based midcareer and recent B.A. programs—appear to have established a reputation for producing "quality" teachers, yet a few program graduates still find that district personnel view nontraditional preparation programs—especially alternative certification programs—somewhat warily. In some cases, local districts have developed a familiarity with these programs that overcomes this wariness. Thus, the success of these nontraditional programs may rest, perhaps to an even greater extent than for traditional teacher preparation programs, on their ability to establish a reputation for producing well-qualified teachers.

In fact, many programs have responded to these factors with changes that have helped them to maintain their enrollments and remain viable. In addition to expanding their recruitment pools, many programs have become institutionalized as part of their affiliated university's regular master's degree or postgraduate teacher certification programs. Some programs initially run by school districts (retraining or alternative certification programs) eventually become university-based as well. They are thus protected from some of the vicissitudes of local funding and labor market shifts, profiting as well from the faculty and programmatic resources of the university. The programs in turn strengthen the capacity of universities to develop and maintain postbaccalaureate teacher education programs, sometimes also encouraging innovation in "regular" program organization and curricula.

The more highly developed nontraditional recruitment programs tend to exist at institutions where innovation in teacher preparation is accepted. For example, some universities that sponsor midcareer programs already have experience with Master of Arts in Teaching (MAT) programs; those sponsoring recent B.A. programs are forging graduate-level "5th-year" models in line with recent reform initiatives, including the recommendations of the Holmes Group and the Carnegie Forum.

RECRUITMENT POOLS TAPPED BY NONTRADITIONAL PROGRAMS

The individuals recruited by the programs we studied differed from "traditional" program recruits in several ways. Overall, they were older, more likely to be male than the overall teaching force (but more likely to be female than the mathematics and science teaching force), and more likely to be members of minority groups. Though minority enrollments in recent B.A. programs in our sample were roughly the same as national norms (about 10 percent), the other program types, including the intense and costly midcareer programs, had substantially higher minority enrollments. Given the current underrepresentation of women and minorities in mathematics and science teaching, the ability of these programs to tap these sectors of the reserve pool is one of their clearest strengths.

Our survey results show that retirees do not appear to constitute a large part of the reserve pool. Programs that initially focused on retirees have usually had to turn to other sources of recruits to maintain their enrollments. Homemakers also appear to participate in these programs in very small numbers. Recent college graduates, on the other hand, enter nontraditional teacher preparation programs in greater numbers, although relatively more of these recruits are likely to choose some other career upon graduation or after entering teaching temporarily.

Teachers from other subject areas are also a good pool from which to draw; however, retraining programs have high attrition rates. Also, after retraining, many of these individuals do not appear to obtain mathematics or science teaching positions, at least within the first year or two. Finally, individuals who change occupations in midcareer also appear to be a smaller, yet viable part of the reserve pool. Programs tapping this source are usually located in areas where there are many relevant occupational pools from which to draw.

Interestingly, a fair number of our sample's "new recruits" to teaching were not in fact brand new entrants to the occupation. About half of the program participants were former teachers; many of them had already been teaching mathematics or science without certification. Of the retrainees, one-third were teaching mathematics or science before they entered a program to "retrain" (and certify) them for this task. Of the nonretrainees, nearly 20 percent had taught previously, virtually all of them in mathematics and science. Most were becoming certified through alternative certification programs, presumably having taught for some period of time without certification. These recruits, then, had already chosen to teach; the program's role was more to help them satisfy minimal requirements than to attract them to the occupation.

Obtaining a sufficient number of recruits from any of these pools requires active program recruitment; individuals typically are not aware of the existence of special postgraduate programs or of sources of financial aid that would allow them to prepare for teaching. Although active recruitment does appear to generate sufficient interest and enrollment, it also has a cost, in that it tends to absorb limited program funding.

Programs targeted toward midcareer transfers and retirees from other science-related occupations have sometimes found that recruitment is made more difficult by teaching's low salary level; many potential candidates lose interest once they discover how much a teacher actually earns. In our sample, relatively few recruits entered teaching from science fields; those who did tended to be from lower-paid occupations, such as technicians and support workers. Programs targeted at midcareer changers seem the most successful at attracting workers from higher-status scientific occupations, but these are also the smallest and most costly programs. Overall, teaching's economic status appears to serve as a strong limitation to recruitment efforts among those in higher-paying occupations.

The difficulty of recruiting individuals from science-related fields should not be surprising, as they earn on average nearly 40 percent more than elementary and secondary school teachers of roughly the same age, education, and experience level. Our analysis of data from the National Science Foundation's registry of scientists and engineers found, for example, that only 0.5 percent of individuals who were working in science fields in 1970 ever switched to elementary or secondary school teaching during the decade of the 1970s. Most of these stayed only a year or two before leaving teaching for other pursuits. For a similar cohort working in science fields in 1980, only 0.2 percent switched to teaching in 1982 or 1984. By comparison, 4 percent of that sample had received degrees in education but were working in occupations other than teaching. Thus, the number of individuals who had "defected" from teaching outnumbered the number of even temporary recruits by 20 to 1 within this scientifically oriented pool.

These analyses suggest that recruitment efforts should focus on those individuals for whom teacher salaries will not be a strong disincentive to enter or remain in teaching. Both our survey and National Science Foundation data show that entrants to teaching from the scientific workforce tend to come from the lower-paying occupations; those with better economic opportunites are clearly more difficult to lure into teaching. Even among those nontraditional recruits who do enter teaching, there is a great deal of dissatisfaction with teacher salaries. Unless and until teaching salaries are made more competitive with other opportunities available to college graduates with mathematics or science training, teacher recruitment efforts will have to continue to rely on the more idealistic attractions of teaching, and on those individuals who are willing to place such idealism above more pragmatic financial concerns.

RECRUITS' PREPARATION EXPERIENCES

Although we have stressed the commonalities among nontraditional teacher preparation programs, the ways in which they differ from one another have the more pronounced influence on recruits' preparation and early teaching experiences. Although all of the programs strive to reduce or overcome some of the potential barriers to entry into teaching, they do so in very different ways. Program duration, intensity, and content vary tremendously, as do financial aid availability and the prospect of quickly assumed paid employment. For example, the extent of required study varies substantially across programs, ranging from as few as nine course credits to as many as 45, conducted over as little as 16 weeks to as much as two or more years, completed before entry into teaching or after a full-time job has been assumed, managed in concert with a guided practicum or without any practicum or student teaching experience at all.

Recruits' comments on their preparation and teaching experiences reflect this variation; they also indicate the challenges facing preparation programs that strive to be both efficient in terms of time and costs and effective in terms of adequate preparation. Programs must find the optimal balance between a number of competing needs and desires: program brevity versus intensity; moving recruits into the classroom quickly versus ensuring that they are fully prepared for the classroom experience; and keeping time and funding demands to a minimum versus providing sufficient practicum support and experience.

In our sample, recruits in midcareer programs—the smaller, more selective and costly programs—were the most satisfied with their coursework and practicum experiences, which tend to be fairly intensive and highly supervised. They were followed fairly closely in satisfaction levels by candidates in retraining programs and recent B.A. programs. Alternative certification recruits were the least satisfied with the amount and quality of preparation and supervision they received. Recruits' ratings of the value of their preparation reflected the kinds of difficulties they experienced when they entered classroom teaching. These differences relate directly to specific features of the preparation experience cited by the recruits.

Alternative certification programs provide less pedagogical coursework than other programs, and also tend not to provide any subject matter coursework or extended practicum experience—these recruits' "practicum" consists of their first year(s) of full-time teaching. Ironically, given that these programs presumably emphasize on-the-job training *in lieu of* standard coursework, the alternative program recruits in our sample received substantially less assistance and supervision than recruits in any of the other types of programs. The demands of taking coursework while also teaching full-time are also a difficulty faced by participants in these programs.

Retraining programs provide the subject-area coursework necessary for endorsement in science or mathematics. Tutoring, individualized instruction, and coursework targeted specifically to teachers are program features that these recruits most appreciate. Although these programs focus on subject matter courses rather than pedagogical courses, candidates often expressed a desire for more subject-specific methods courses, for a guided practicum experience, and for opportunities to observe other veteran teachers in these fields. Even experienced teachers feel they have more pedagogical knowledge to acquire when they switch to a new teaching field. Like alternative certification recruits, retrainees often find the dual demands of full-time teaching and intensive coursework difficult to sustain.

Midcareer and recent B.A. programs are similar in that they tend to provide the coursework and practicum experience necessary for full certification and a master's degree in education; recruits sometimes have the option of either obtaining the advanced degree, or merely taking the coursework required for certification. In either case, though, the breadth and depth of required coursework tends to be substantially greater than that offered in alternative certification or retraining programs. These programs also tend to be more selective in their admissions requirements than other nontraditional programs; for those leading to a master's degree, entry requirements are usually based on graduate school requirements. The smaller midcareer programs appear to provide the best practicum experiences, in terms of support and supervision. Some recent B.A. recruits, particularly in those programs that put candidates into the classroom quickly and with less intense supervision, are vocal in requesting more and better practicum support.

In some respects, recruits echoed common themes across program types. Like traditional teacher education students, many participants and graduates tended to feel that their programs overemphasized theory at the expense of the practical knowledge and skills necessary for effective teaching, especially classroom management and organizational/time management skills. They also often felt that they knew less than they would have liked about student motivation, adolescent behavior, and student learning difficulties. Those who were taught only specific teaching techniques without a basis for diagnosing and modifying their teaching approach were caught short when the techniques did not always work.

Many candidates felt that practicing classroom teachers should be more involved in offering coursework as well as supervision, that instructors not themselves familiar with the demands of K–12 teaching were less likely to understand the applications of course knowledge. A frequent comment was that courses needed to better link theory with practice and content with pedagogy.

Programs tend to be designed on the assumption that candidates come either with adequate subject matter background or, in the case of retraining programs, with adequate pedagogical background. In the cause of efficient preparation, they strive to provide additional coursework only in the areas not presumably already mastered. However, candidates in all types of programs often felt that they would have benefited from more coursework in both areas, filling in subject matter gaps or holes in pedagogical understanding—and in the area bridging subject matter to pedagogy, the acquisition of subject-specific teaching methods. It may be that preparation for the complexity of teaching cannot treat content and pedagogy as entirely separate from one another. It is in the merger of content and method, of theory and application, that teaching is better understood.

Perhaps it is because of this need for integration that the teaching practicum experience appears to be the most critical aspect of training provided by these programs. Recruits who had the benefit of practicum experiences were united in citing its importance. Their suggestions for improvements in this aspect of preparation only confirmed its value—increasing the length of short practicums, adding a practicum and observation period to retraining and alternative certification programs that currently have none, ensuring adequate support and supervision, improving selection of mentor teachers, providing opportunities for discussion among recruits and between recruits and current teachers, and increasing the range of classes observed or taught so that candidates are familiar with a wider variety of students and types of classes.

Unfortunately, these types of improvements are often at odds with the desire to keep program time and costs at a minimum. Programs need to be innovative in developing procedures to provide adequate preservice teaching experiences. For example, the use of videotapes and structured cohort experiences and discussions, occasionally used by some of these programs, appear to be two cost-efficient ways of maximizing the value of available practicum experiences. These procedures should not, however, replace the necessary experience of an adequately supervised and appropriately structured teaching practicum.

Most of these programs, particularly the midcareer and recent B.A. programs, are more similar in content to traditional teacher training programs than they are different. However, as new, evolving programs, these nontraditional preparation programs provide (with sufficient financial support) an opportunity to experiment with preparation methods that may improve upon those traditionally used. We would recommend that these programs continue to pursue such innovations, while closely monitoring recruits' reactions to these changes, and their effects on teacher practice and retention.

Appendix

DEFINING THE RESERVE POOL FOR MATHEMATICS AND SCIENCE TEACHERS

It is becoming increasingly clear that the pool of traditional recruits to teaching is insufficient to meet current and future demand. Thus, many positions will need to be filled from other sources. Definitions of teacher supply, therefore, need to be extended to include different pools and modes of entry into teaching. For example, to enlarge the pool of mathematics and science teachers, we need to include in the category of *prospective* teachers, current teachers not certified in mathematics and science, college undergraduates not majoring in education, those with mathematics and science degrees working in other fields, and retirees with requisite college majors.

Like other components of teacher supply and demand, it is undoubtedly true that the size, composition, and potential availability of the reserve pool varies among locations and different fields depending on the characteristics of the local population (age, education, and employment) and economy, the degree of transiency in the population, and current and past school personnel practices. Districts that laid off teachers in the late 1970s and early 1980s, for example, are still rehiring members of their former teaching staffs. Districts with lenient hiring practices may find it easier to tap the reserve pool than those that apply stringent certification, education, or testing requirements. States and districts with more attractive teaching conditions—higher salaries, better working conditions, good benefits—and those that can offer full credit on the salary scale for prior teaching experience may tap larger sectors of the reserve pool. The existence of training programs and financial incentives targeted at encouraging and easing entry to teaching for nontraditional entrants also make a difference in access to various segments of the reserve pool, as we have seen in this report.

The analyses of nontraditional teacher preparation programs reported here have told us a great deal about who enters these programs and about their teaching careers. However, these data are insufficient for defining the reserve pool for mathematics or science teaching, or for estimating the potential size and composition of this pool. To do this, other data sources must be examined. The National Science Foundation maintains a longitudinal dataset on individuals with scientific work experience that provides a useful place to begin an examination of the reserve pool for mathematics and science teaching. In this appendix, we summarize the results of analyses of this dataset, which define three potential reserve pools to teaching from the total scientific workforce, and compare these groups on a number of sociodemographic variables. These analyses help shed light on what subgroups among the scientific workforce are the most productive reserve pools for mathematics and science teaching. This information, in turn, should help policymakers target recruitment efforts to those groups most likely to consider entering teaching.

NSF SURVEYS OF SCIENTISTS AND ENGINEERS

The National Science Foundation's longitudinal registry of scientists and engineers, the Survey of Scientists and Engineers (SSE), provides information on the personal characteristics, education, and current and previous employment of a sample of those identified as scientists and engineers at the time of the decennial census. The sample is drawn from the census, with follow-up surveys conducted in two to three year cycles over the course of the decade. The analyses discussed here are based on the sample drawn from the 1980 census, with surveys conducted in 1982 and 1984. The 1980s sample was selected from the census sample in two stages, using differential sampling rates within each of 51 occupational strata identified in the first stage. The 1982 SSE sample includes the following "target" occupations: operations and computer specialists, engineers, mathematical specialists, life scientists, physical scientists, environmental scientists, psychologists, social scientists, college teachers not identified as teaching in one of the above fields, managers and administrators not elsewhere classified, and the labor reserve. The sample for 1984 was selected from the respondents of the 1982 survey. Those sample individuals classified as scientists or engineers were selected with certainty; others were sampled at a 70 percent rate. Response rates (unadjusted) of 72.1 and 71.8 percent were obtained in 1982 and 1984. The 1984 sample consisted of approximately 56,000 respondents. For our analyses, we omitted all individuals whose census occupation was not specifically listed as science-related (e.g., "not otherwise classified" categories); this reduced our sample size to approximately 49,000.

Defining Reserve Pools

Defining all individuals surveyed by the SSE as the reserve pool for mathematics and science teaching is not very plausible. Using the information gathered from our analyses of nontraditional programs and our theoretical framework, we defined three subgroups that could plausibly be expected to have a somewhat higher propensity to enter teaching. These subgroups are defined, using census (1980) data, as those:

- Working in the service sector. The service sector includes elementary and secondary schools, colleges and universities, business, trade, and vocational schools, libraries, job training and vocational rehabilitation services, child day care services, residential care facilities, museums, galleries and zoos, and other educational and social services.
- With an education degree.
- Who listed their primary work activity as "teaching and training."

Our rational for selecting these three subgroups is as follows: First, one can argue that individuals working in the service sector are either in environments that would facilitate transitions to teaching, or have jobs that involve "serving" people to some degree; those with a service orientation are more likely to choose to enter teaching than are those without such an orientation. Second, individuals with education degrees, at least at some point, had considered teaching. One could, of course, make exactly the opposite point—that these individuals, having considered the education field, have self-selected *out* of the field and, therefore, are unlikely to return. Either way, it is important to examine this group to see what occupational choices they have made and to estimate their propensity to enter teaching from other occupational endeavors. The third subgroup is perhaps the least ambiguous. Our data showed that most nontraditional recruits had had some prior teaching experience, both in schools and in private industry.[1] It seems plausible that individuals who have selected into jobs involving teaching or training might well be persuaded to become teachers. It should be noted that there is some degree of overlap between these three subgroups. The service sector and those whose work primarily involves teaching or training have a particularly high degree of overlap, as both subgroups include postsecondary teachers.

---

[1]This is probably due to selection on both sides. That is, individuals who choose to teach in other settings may be more likely to enter elementary or secondary teaching, *and* programs are likely to select for teacher preparation individuals with some type of teaching experience.

Table A.1 describes the entire sample and the three subgroups in terms of demographic variables and career backgrounds. The service sector subgroup includes 9.3 percent of the total sample; the education and teaching/training subgroups include 3.7 and 11.0 percent, respectively. Where we have data, we show the propensity of these individuals to enter teaching. Unfortunately, the sample drawn in 1980 deliberately excluded precollege teachers; moreover, the 1984 sample undersampled those not classified as scientists and engineers in 1982. Data on teachers and moves into teaching are thus very sparse, and probably underestimate to some unknown degree movement into (and out of) teaching.

With this caveat in mind, we see that members of these reserve pool subgroups are indeed more likely than scientists in general to enter teaching, although the actual proportions are very small (no more than 1 percent for any of the subgroups). In fact, of the total pool of scientists, only 52 ever entered K-12 teaching in 1982 or 1984. Further, 92.3 percent of these individuals came from occupations that primarily involve teaching or training; one-third transferred from each of the other two subgroups. Thus, these subgroups—particularly the subgroup with teaching experience—have both theoretical and empirical support for being viable sectors of the reserve pool from the scientific workforce.

Looking at the proportion within each group who were elementary and secondary teachers in 1982 or 1984, and comparing this to the proportion who were teachers in *both* years, reveals considerable movement. About 1 percent of each subgroup were classified as teachers in one year or the other, with 0.2 to 0.8 percent teaching in any one year. However, less than 0.1 percent of each subgroup were teachers in *both* years. Thus, even among the small number of individuals who do transfer into teaching, most leave teaching within the first four years. These data reveal the teaching career histories of these reserve pool groups; we turn now to an examination of the demographic, educational, and science career histories of these groups.

First, the subgroups differ from the science workforce as a whole in gender composition; women constitute roughly one-quarter of each subgroup, but only one-tenth of the overall sample. There is little difference, however, by race or age. When one looks at years of education completed beyond high school, it is interesting to note that individuals in the service sector and teaching/training subgroups tend to have higher levels of educational attainment. This is probably attributable to the fact that all or most postsecondary instructors are included in each of these subgroups.

Table A.1

SELECTED CHARACTERISTICS OF POTENTIAL RESERVE POOLS FOR MATHEMATICS AND SCIENCE TEACHERS, BY SUBGROUP

Selected Characteristics[a]	Total Sample	Service Sector	Education Degrees	Teaching/ Training
Percent entering teaching in 1982 or 1984	0.2	0.9	0.9	1.0
Percent teaching in 1982	0.1	0.2	0.5	0.4
Percent teaching in 1984	0.1	0.8	0.5	0.8
Percent teaching in both years	<0.1	<0.1	<0.1	<0.1
Sex				
Percent male	89.9	75.4	72.6	80.3
Percent female	10.1	24.6	27.4	19.7
Race				
Percent Asian/Pacific Islander	5.4	3.8	2.5	3.3
Percent black	2.2	2.3	4.3	3.1
Percent white	91.0	92.0	91.8	92.2
Percent other	1.5	1.9	1.3	1.3
Age (years)				
Mean	42.2	42.1	44.4	42.1
Median	39.0	40.0	41.0	40.0
Range	17–92	20–83	24–86	20–83
Years of education (beyond high school)				
Mean	5.3	6.7	5.8	6.6
Median	5.0	7.0	6.0	7.0
Range	0–15	0–15	1–15	0–15
(N)[b]	(49,160)	(4,578)	(1,820)	(5,405)
Occupation in 1980				
Engineers, architects, and surveyors	62.5	7.9	37.4	27.7
Mathematical and computer scientists	8.9	4.8	13.8	8.2
Natural scientists	12.3	12.9	2.4	11.5
Health diagnosing occupations	0.0	0.0	1.3	0.0
College teachers	3.9	50.1	2.9	31.0
Other teachers	0.0	0.0	7.3	0.0
Educational and vocational counselors	0.0	21.3	0.0	0.0
Librarians, archivists and curators	0.0	0.0	8.5	0.0
Social scientists and urban planners	6.6	0.0	26.0	16.4
Computer programmers	5.8	3.0	14.1	5.3
Median 1982 salary	$33,000	$31,298	$30,000	$31,000
Median 1984 salary	$37,900	$35,000	$34,000	$35,000

[a] Based on 1980 census data, unless otherwise noted.
[b] Unweighted number of cases. The proportions and averages shown are based on weighted frequencies.

Well over three-fifths of the total sample were engineers, architects, or surveyors in 1980. The subgroups tend to display somewhat different occupational patterns. Overall, fewer subgroup members were engineers; more were college instructors, counselors, or social scientists. Obviously, this difference reflects, at least for the service sector and teaching/training groups, our definitional criteria for these subgroups. It is interesting to note, however, that the education subgroup—the group that can be considered "defectors" from precollege teaching—is most commonly found in engineering occupations. Although it is not surprising to find that many of those with education degrees are also social scientists, it is somewhat surprising to see that a relatively large proportion are in computer-related occupations. It appears that the education "defectors" who move into scientific fields are often drawn to high-demand occupations.

Finally, the salary data lend support to our hypothesis that those with lower opportunity costs may form a potentially more productive reserve pool. Our three subgroups do indeed have lower median salaries in both years, in spite of the fact that they tend to be more highly educated.

Characteristics of Elementary/Secondary School Teachers

We also analyzed data for the subsample of 52 individuals who were elementary/secondary school teachers in either 1982 or 1984. This sample size is extremely small; nevertheless, the profile painted by these data is worth examining. Tables 4.3 and 4.4 in the main text present the results of this analysis; the results are also discussed in the main text. Summarizing those results, we found that our sample of teachers is disproportionately female, with a higher proportion of minorities. Their educational attainment is high; over three-fifths have at least a master's degree (Table 4.3). Again, the salary data for this group indicate that those who enter teaching earn less in science-related careers than does the sample as a whole,[2] and earn even less when they are employed as teachers (Table 4.4).

---

[2] Those entering teaching are also less well-paid, on average, than each of the subgroups, suggesting that even within reserve pool subgroups, it is the lower-paid individuals who transfer to teaching.

Conclusions

We have attempted here to delineate, from an overall sample of scientists and engineers, subgroups that might prove to be potentially more productive reserve pools for mathematics and science teachers. The propensity to enter teaching is extremely small (no more than 1 percent) among any group; it is, however, higher in our subgroups than in the overall sample, giving some credence to our selection criteria. We defined these groups as those who, in the base census survey: (a) were working in the service sector, (b) had an education degree, and (c) listed "teaching or training" as their primary work activity. Overall, these groups tend to be (relative to the science workforce as a whole), female, highly educated, poorly paid, and in the social science occupations.

The sample of elementary/secondary teachers drawn from the larger scientific reserve pool was far too small to support detailed analysis. However, we find that this group, in many ways similar to our three more focused reserve pools, tended to be disproportionately female, to have higher proportions of minorities, to be highly educated, poorly paid, to underrepresent engineers, but overrepresent biological, physical, mathematical, and social scientists, and to have a high degree of occupational mobility.

From these results, it appears that elementary and secondary teaching has historically drawn extremely few "recruits" from the scientific workforce, even from among those sectors of this workforce that might be expected to have (and do in fact have) a greater propensity for making this lateral career move. These reserve pools—and those who do enter teaching, predominantly from these reserve pools—are disproportionately composed of the demographic and occupational groups that tend to be less well-paid (even though they are *not* less well-educated). It thus appears that economic factors are most crucial in determining a science-field worker's propensity to enter teaching.

Clearly, however, if policymakers wish to tap the science-related workforce as a reserve pool for mathematics and science teaching, we need more information on this potential pool. For example, a sample from the SSE could be surveyed to determine what proportion of scientists would be willing to consider a lateral career move into teaching under various circumstances, including increased access to nontraditional route programs, increased salary schedules for teaching, etc. In addition to helping clarify the size and composition of this sector of the teaching reserve pool, a special survey could also (1) shed light on what policies are likely to be most effective in attracting members of this pool; (2) reveal the nature of training these individuals have (and therefore what training they would need), including the recency and breadth of their subject-area knowledge; and (3) what policies or other changes in teaching might have convinced these individuals to enter (or remain in) teaching when they were college students (or graduates). This type of information would greatly aid efforts to recruit elementary and secondary teachers from this potentially valuable sector of the reserve pool.

BIBLIOGRAPHY

Adams, R. D., "Teacher Development: A Look at Changes in Teacher Perceptions and Behavior Across Time," *Journal of Teacher Education,* Vol. 33, No. 4, 1982, pp. 40–43.

Adelman, N. E., *An Exploratory Study of Teacher Alternative Certification and Retraining Programs,* Policy Studies Associates, Washington, D.C., October 1986.

American Association of Colleges for Teacher Education (AACTE), *Teacher Education Policy in the States: 50-State Survey of Legislative and Administrative Actions,* Washington, D.C., 1985.

American Association of Colleges for Teacher Education (AACTE), *Teacher Education Policy in the States: 50-State Survey of Legislative and Administrative Actions,* Washington, D.C., 1987.

American Federation of Teachers (AFT), *Survey and Analysis of Salary Trends,* Washington, D.C., July 1988.

Applegate, J. H., and T. J. Lasley, "Student's Expectations for Early Field Experience," *Texas Tech Journal of Education,* Vol. 12, No. 1, 1985, pp. 27–36.

Association for School, College, and University Staffing (ASCUS), *Teacher Supply/Demand 1986,* ASCUS, Madison, Wisconsin, 1986.

Bills, R. E., V. M. Macagnoni, and R. J. Elliot, "Student Teacher Personality Change as a Function of the Personalities of Supervising and Cooperating Teachers," *Journal of Teacher Education,* Vol. 17, 1966, pp. 245–246.

Boyer, E. L., *High School: A Report on Secondary Education in America,* Harper and Row, New York, 1983.

Brown, M., and A. Williams, "Lifeboat Ethics and the First-Year Teacher," *The Clearinghouse,* Vol. 51, 1977, pp. 73–75.

Capper, J., *A Study of Certified Teacher Availability in the States,* Council of Chief State School Officers, Washington, D.C., 1987.

Carey, N., B. Mittman, and L. Darling-Hammond, *Recruiting Mathematics and Science Teachers Through Nontraditional Programs: A Survey,* The RAND Corporation, N-2736-FF/CSTP, May 1988.

Carnegie Forum on Education and the Economy, *A Nation Prepared: Teachers for the 21st Century,* Washington, D.C., 1986.

Carroll, C. D., *High School and Beyond Tabulation: Background Characteristics of High School Teachers,* National Center for Education Statistics, Washington, D.C., 1985.

Center for Education Statistics (CES), "Survey of Recent College Graduates," unpublished summary tables, Washington, D.C., May 1986.

Center for Education Statistics (CES), *The Digest of Education Statistics*, Washington, D.C., 1987a.

Center for Education Statistics (CES), "Public School Teacher Perspectives on School Discipline," *OERI Bulletin*, CS87-387, Office of Educational Research and Improvement, U.S. Department of Education, October 1987b.

Chapman, D., "A Model of the Influences on Teacher Retention," *Journal of Teacher Education*, Vol. 34, No. 5, 1983, pp. 43–49.

Coley, R. J., and M. E. Thorpe, *Responding to the Crisis in Math and Science Teaching: Four Initiatives*, Educational Testing Service, Princeton, New Jersey, 1985a.

Coley, R. J., and M. E. Thorpe, *A Look at the MAT Model of Teacher Education and Its Graduates: Lessons for Today*, Educational Testing Service, Princeton, New Jersey, 1985b.

Darling-Hammond, L., *Beyond the Commission Reports: The Coming Crisis in Teaching*, The RAND Corporation, July 1984.

Darling-Hammond, L., and L. Hudson, *Precollege Science and Mathematics Teachers: Supply, Demand, and Quality*, paper prepared for the National Science Foundation, 1987.

Darling-Hammond, L., and B. Berry, *The Evolution of Teacher Policy*, The RAND Corporation, JRE-01, March 1988.

Darling-Hammond, L., K. J. Pittman, and C. Ottinger, "Career Choices for Minorities: Who Will Teach?" paper prepared for the National Education Association and Council of Chief State School Officers, October 1987.

Elias, P., M. L. Fisher, and R. Simon, *Helping Beginning Teachers Through the First Year: A Review of the Literature*, Educational Testing Service, Princeton, New Jersey, 1980.

Feistritzer, E. C., *The Making of a Teacher*, National Center for Educational Information, Washington, D.C., 1984.

Feistritzer, E. C, *Profile of Teachers in the U.S.*, National Center for Educational Information, Washington, D.C., 1986.

Fox, S. R., *Scientists in the Classroom: Two Strategies*, National Institute for Work and Learning, Washington, D.C., 1986.

Gaede, O., "Reality Shock: A Problem Among First-Year Teachers," *The Clearinghouse*, Vol. 51, 1978, pp. 405–409.

Gallup, A., "The Gallup Poll of Teachers' Attitudes Toward the Public Schools," *Phi Delta Kappan*, Vol. 66, No. 2, 1984, pp. 97–107.

Grissmer, D. W., and S. Kirby, *Teacher Attrition: The Uphill Climb To Staff the Nation's Schools*, The RAND Corporation, R-3512-CSTP, August 1987.

Howe, T. G., and J. A. Gerlovich, "National Study of the Estimated Supply and Demand of Secondary Science and Mathematics Teachers," working document, Iowa State University, Ames, Iowa, 1982.

Hudson, L., S. N. Kirby, N. Carey, B. Mittman, and B. Berry, *Recruiting Mathematics and Science Teachers Through Nontraditional Programs: Case Studies*, The RAND Corporation, N-2768-FF/CSTP, August 1988.

Johnson, R. T., and D. W. Johnson, "Student-Student Interaction: Ignored but Powerful," *Journal of Teacher Education*, Vol. 36, No. 4, 1985, pp. 22–26.

Lippman, S. A., and J. J. McCall (eds.), *Studies in the Economics of Search*, North-Holland, Amsterdam, 1979.

Loadman, W. E., S. M. Brookhart, and S. Wongwanich, "Perceptions and Attitudes about Teaching and Teacher Education," paper presented at the annual meeting of the American Educational Research Association, New Orleans, Louisiana, April 1988.

Marso, R. N., and F. L. Pigge, "Differences Between Self Perceived Job Expectations and Job Realities of Beginning Teachers," *Journal of Teacher Education*, Vol. 38, No. 6, 1987, pp. 53–56.

McLaughlin, M. W., R. S. Pfeifer, D. Swanson-Owens, and S. Yee, "Why Teachers Won't Teach," *Phi Delta Kappan*, Vol. 67, No. 6, 1986, pp. 420–426.

Medley, D. M., "Experiences with the OScAR Technique," *Journal of Teacher Education*, Vol. 14, 1963, pp. 273–276.

Metropolitan Life, *The Metropolitan Life Survey of the American Teacher 1986: Restructuring the Teaching Profession*, New York, 1986.

Metzner, S., W. A. Nelson, and R. M. Sharp, "On-Site Teaching: Antidote for Reality Shock," *Journal of Teacher Education*, Vol. 33, No. 2, 1972, pp. 194–198.

Murnane, R. J., and R. J. Olsen, "Factors Affecting Length of Stay in Teaching: Evidence from North Carolina," paper presented at the annual meeting of the American Educational Research Association, New Orleans, Louisiana, April 1988.

National Center for Education Statistics (NCES), *The Condition of Education*, U.S. Department of Education, Washington, D.C., 1983.

National Commission on Excellence in Education, *A Nation At Risk*, U.S. Government Printing Office, Washington, D.C., 1983.

National Education Association (NEA), *Status of the American Public School Teacher: 1985-86*, Washington, D.C., 1987.

National Science Board, *Science and Engineering Indicators—1987*, U.S. Government Printing Office, Washington, D.C., 1987.

National Science Board Commission on Precollege Education in Mathematics, Science, and Technology, *Educating Americans for the 21st Century*, Washington, D.C., 1983.

National Science Teachers Association, "Survey of College and University Placement Officers," *Science Indicators: The 1985 Report*, National Science Board, Washington, D.C., 1982.

Natriello, G., K. Zumwalt, A. Hansen, and A. Frisch, *Characteristics of Entering Teachers in New Jersey*, unpublished paper, Teachers College, Columbia University, New York, June 1988.

Nelson, P., "Information and Consumer Behavior," *Journal of Political Economy*, Vol. 78, 1970, pp. 311–329.

O'Rourke, B., "'Lion Tamers and Baby Sitters': First-Year English Teachers' Perceptions of Their Undergraduate Teacher Preparation," *English Education*, Vol. 15, No. 1, 1983, pp. 17–24.

Pencavel, J., "Wages, Specific Training and Labor Turnover in U.S. Manufacturing Industries," *International Economic Review*, Vol. 13, 1972, pp. 53–54.

Pigge, F. L., "Teacher Competencies: Need, Proficiency, and Where Proficiency Was Developed," *Journal of Teacher Education*, Vol. 29, No. 4, 1978, pp. 70–75.

Plisko, V. W., "Teacher Preparation," in V. W. Plisko (ed.), *The Condition of Education*, 1983 ed., U.S. Government Printing Office, Washington, D.C., 1983.

Rosenholtz, S. J., D. Bassler, and K. Hoover-Dempsey, "Organizational Inducements for Teaching," interim report submitted to the National Institute of Education, University of Illinois, 1985.

Rumberger, Russell, "The Shortage of Mathematics and Science Teachers: A Review of the Evidence," *Educational Evaluation and Policy Analysis*, Vol. 7, No. 4, 1985, pp. 355–369

Ryan, K., J. Applegate, V. R. Flora, J. Johnston, T. Lasely, G. Mager, and K. Newman, "'My Teacher Education Program? Well...': First-Year Teachers Reflect and React," *Peabody Journal of Education*, Vol. 56, 1979, pp. 267–271.

Ryan, K., K. K. Newman, G. Mager, J. Applegate, T. Lasely, R. Flora, and J. Johnston, *Biting the Apple: Accounts of First-Year Teachers*, Longman, New York, 1980.

Scannell, M., "State Characteristics Associated with Policies Restricting Entry to Teaching," paper presented at the American Educational Research Association Annual Meeting, April 1988.

Sedlak, M., and S. L. Schlossman, *Who Will Teach? Historical Perspectives on the Changing Appeal of Teaching As a Profession*, The RAND Corporation, R-3472-CSTP, November 1986.

Shymansky, J., and B. Aldridge, "The Teacher Crisis in Secondary School Science and Mathematics," *Educational Leadership*, No. 40, 1982, pp. 61–62.

Stolzenberg, R. M., and R. Giarrusso, "When Students Make the Selections: How the MBA Class of '87 Picked Its Schools," *Selections: The Magazine of the Graduate Management Admission Council*, Vol. 4, No. 2, Autumn 1987, pp. 14–20.

Stolzenberg, R. M., J. Abowd, and R. Giarrusso, "Abandoning the Myth of the Modern MBA Student," *Selections: The Magazine of the Graduate Management Admission Council*, Vol. 3, No. 2, Autumn 1986, pp. 9–21.

U.S. Bureau of the Census, *Statistical Abstract of the United States: 1988* (108th ed.), Washington, D.C., 1987.

Veenman, S., "Perceived Problems of Beginning Teachers," *Review of Educational Research*, Vol. 54, No. 2, 1984, pp. 143–178.

Vetter, B. M., and E. L. Babco, *Professional Women and Minorities: A Manpower Data Resource Service*, 7th ed., Commission on Professionals in Science and Technology, 1987.

Weiss, I. R., *Report of the 1985–86 National Survey of Science and Mathematics Education*, Research Triangle Institute, Research Triangle Park, North Carolina, 1987.

Wise, A. E., L. Darling-Hammond, and B. Berry, *Effective Teacher Selection: From Recruitment to Retention*, The RAND Corporation, R-3462-NIE/CSTP, January 1987.

Zabalza, A., "The Determinants of Teacher Supply," *The Review of Economic Studies*, Vol. 46, No. 142, 1979, pp. 131–147.

Ministry of Education, Ontario
Information Services & Resources Unit,
13th Floor, Mowat Block, Queen's Park,
Toronto   M7A 1L2